ENERGY POLICIES, POLITICS AND PRICES

REDUCED ENERGY USE THROUGH DEMAND RESPONSE

SELECT EXPERIENCES FROM THE U.S. AND ABROAD

ENERGY POLICIES, POLITICS AND PRICES

Additional books in this series can be found on Nova's website under the Series tab.

Additional E-books in this series can be found on Nova's website under the E-book tab.

Energy Policies, Politics and Prices

Reduced Energy Use through Demand Response

Select Experiences from the U.S. and Abroad

Peter N. Ryan
Editor

nova publishers
New York

For permission to use material from this book please contact us:
Telephone 631-231-7269; Fax 631-231-8175
Web Site: http://www.novapublishers.com

Library of Congress Cataloging-in-Publication Data

ISBN: 978-1-62948-074-9

Published by Nova Science Publishers, Inc. † New York

Contents

Preface		**vii**
Chapter 1	Addressing Energy Demand through Demand Response: International Experiences and Practices *Bo Shen, Girish Ghatikar, Chun Chun Ni, Junqiao Dudley, Phil Martin and Greg Wikler*	**1**
Chapter 2	Demand Response in U.S. Electricity Markets: Empirical Evidence *Peter Cappers, Charles Goldman and David Kathan*	**49**
Chapter 3	Market and Policy Barriers for Demand Response Providing Ancillary Services in U.S. Markets *Peter Cappers, Jason MacDonald and Charles Goldman*	**75**
Index		**141**

PREFACE

Demand response (DR) is a load management tool which provides a cost-effective alternative to traditional supply-side solutions to address the growing demand during times of peak electrical load. According to the U.S. Department of Energy (DOE), demand response reflects "changes in electric usage by end-use customers from their normal consumption patterns in response to changes in the price of electricity over time, or to incentive payments designed to induce lower electricity use at times of high wholesale market prices or when system reliability is jeopardized." The California Energy Commission (CEC) defines DR as "a reduction in customers' electricity consumption over a given time interval relative to what would otherwise occur in response to a price signal, other financial incentives, or a reliability signal." This latter definition is perhaps most reflective of how DR is understood and implemented today in countries such as the U.S., Canada, and Australia where DR is primarily a dispatchable resource responding to signals from utilities, grid operators, and/or load aggregators (or DR providers). This book examines select experiences from the U.S. and abroad corresponding to reduced energy usage through demand response.

Chapter 1 – Report of Environmental Energy Technologies Division, Lawrence Berkeley National Laboratory on Addressing Energy Demand through Demand Response: International Experiences and Practices, dated June 2012.

Chapter 2 – Report of a Berkeley National Laboratory on Demand Response in U.S. Electricity Markets: Empirical Evidence, dated June 2009.

Chapter 3 - Report of the Lawrence Berkeley National Laboratory on Market and Policy Barriers for Demand Response Providing Ancillary Services in U.S. Markets, dated March 2013.

In: Reduced Energy …
Editor: Peter N. Ryan

ISBN: 978-1-62948-074-9

Chapter 1

ADDRESSING ENERGY DEMAND THROUGH DEMAND RESPONSE: INTERNATIONAL EXPERIENCES AND PRACTICES*

Bo Shen, Girish Ghatikar, Chun Chun Ni, Junqiao Dudley, Phil Martin and Greg Wikler

LIST OF ACRONYMS

AESO	Alberta Energy System Operator
AMI	advanced metering infrastructure
AMP	aggregator managed portfolio
Auto-DR	automated demand response
BRA	base residual auction
C&I	commercial and industrial
CAP	Climate Action Plan
CEC	California Energy Commission
CPP	critical peak pricing
CSPs	curtailment service providers
DECC	U.K. Department of Energy and Climate Change
DLC	direct load control

* This is an edited, reformatted and augmented version of Environmental Energy Technologies Division, Lawrence Berkeley National Laboratory, Publication No. LBNL-5580E, dated June 2012.

DOE	U.S. Department of Energy
DPCR5	distribution price control review
DR	demand response
DRAS	demand response automation server
DRRC	Demand Response Research Center
DSM	demand side management
EE	energy efficiency
EEPS	Energy Efficiency Portfolio Standard
EILS	emergency interruptible load service
EISA 2007	Energy Independence and Security Act of 2007
EMCS	energy management control systems
EPACT	The Energy Policy Act of 2005
ERCOT	Electric Reliability Council of Texas
FCM	forward capacity market
FERC	U.S. Federal Energy Regulatory Commission
FRCC	Florida Reliability Coordinating Council
HVAC	heating ventilation and air-conditioning
I/C	Interruptible/curtailable
IA	incremental auctions
IESO	independent electricity system operator
IOUs	investor-owned utilities
IT	information technology
LMP	locational marginal price
M&V	measurement and verification
MRO	Midwest Reliability Organization
NOCs	network operation centers
NPCC	Northeast Power Coordinating Council, Inc.
OPA	Ontario Power Authority
PG&E	California Pacific Gas and Electric
PPA	power purchase agreement
PTR	peak time rebates
RFC	ReliabilityFirst Corporation
RPM	reliability pricing model
RTDR	real time demand response
RTEG	real time emergency generation
RTO	regional transmission organization
RTP	real-time pricing
SCE	Southern California Edison
SERC	Southeastern Electric Reliability Council

SPP	Southwest Power Pool
SRM	sycnhronized reserve market
STOR	short term operating reserves
TOU	time-of-use
TVA	Tennessee Valley Authority
WECC	Western Electricity Coordinating Council
WEM	whosesale electricity market

1. INTRODUCTION

1.1. Definition of Demand Response

Demand response (DR) is a load management tool which provides a cost-effective alternative to traditional supply-side solutions to address the growing demand during times of peak electrical load. According to the US Department of Energy (DOE), demand response reflects "changes in electric usage by end-use customers from their normal consumption patterns in response to changes in the price of electricity over time, or to incentive payments designed to induce lower electricity use at times of high wholesale market prices or when system reliability is jeopardized." [1] The California Energy Commission (CEC) defines DR as "a reduction in customers' electricity consumption over a given time interval relative to what would otherwise occur in response to a price signal, other financial incentives, or a reliability signal." [2] This latter definition is perhaps most reflective of how DR is understood and implemented today in countries such as the US, Canada, and Australia where DR is primarily a dispatchable resource responding to signals from utilities, grid operators, and/or load aggregators (or DR providers).

1.2. Benefits Brought by Demand Response

There are a variety of benefits brought by DR, ranging from the environmental to the economic.

Environmental Benefits

By reducing electric demand to ensure the sufficiency of existing supply, rather than increasing supply to meet rising demand, DR avoids power plant operation and its associated emissions. Moreover, because DR capacity is distributed, there are added benefits due to the avoidance of electrical losses in the transmission and distribution lines typically experienced from centrally-generated utility power. The US-based energy consultancy Synapse Energy Economics addressed this issue in its study of DR and air emissions in the US market of ISO New England:

> "...when DR operates it reduces system line losses relative to reference case system operation. This is because when energy is provided to customers from the grid it often comes from power plants a considerable distance from the point of end use, and energy is lost in transmission. Usually line losses are in the range of 5 to 10 percent, but they can be higher during periods when transmission lines are heavily loaded. In contrast, the DR resource – be it a load reduction or a generator – is located at the site of energy use, so no energy is lost in transmission."
> "Because DR avoids line losses, a DR resources of five MW is comparable to a grid-connected asset of slightly larger than five MW...For example, a five-MW DR resource might be credited as providing 5.5 MW of reserve capacity is average system line losses during DR events were determined to be roughly 10 percent."[3]

In addition, because DR is often procured on a forward basis, it may not only offset the operation of power plants but also their very construction. In this manner, the environmental benefits of DR extend to the avoided emissions associated with the construction of the materials for the power plant itself (i.e. cement, steel, etc.), as well as the potential ecological impact that may have resulted should the unit have been constructed.

The use of DR for non- peak-shaving purposes such as for ancillary services, also comes with significant environmental benefits, despite the very short duration dispatches of such resources. In many systems, ancillary services (also known as reserves), are primarily provided by plants in running operating mode, as there may be an insufficient number of quick-start generating units able to start, synchronize, and export power to the grid in the requisite period of time. These plants tend to be fueled by diesel or oil, which add to local and regional pollution. Increased use of quick-response DR can reduce the need for power plants to run in operating mode, as well as potentially lead to a more efficient overall use of resources within the system.

Economic Benefits

The economic benefits of DR oftentimes may be more significant than the environmental benefits. While there is a clear environmental benefit to avoiding or reducing power plant operation, the targeted usage of DR will not save the same amount of energy as permanent load reductions that come from energy efficiency measures. As peak periods are relatively infrequent, so too tends to be the use of DR. Yet, the infrequent spikes in demand have a significant economic impact: in many systems, 10% (or more) of costs are incurred to meet demands which occur less than 1% of the time.[4] Reducing this peak demand through DR programs means that the capacity requirements which drive investments in generation, transmission, and distribution assets can also be proportionally reduced. The US-based energy consultancy the Brattle Group, in its 2007 paper "The Power of Five Percent," found that a 5% reduction in peak demand would have resulted in avoided generation and T&D capacity costs of $2.7 billion per year.[5]

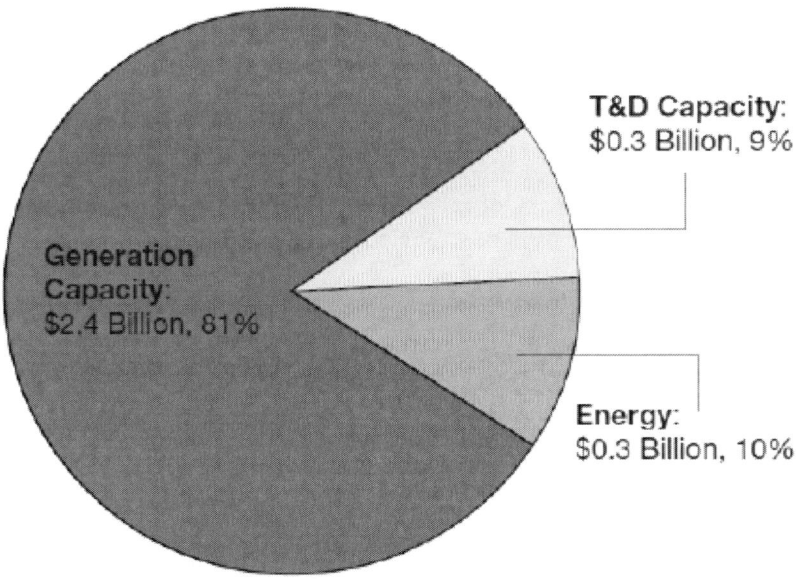

Figure 1. Annual Benefits of 5% Remand Response in the US.[6]

In addition, the use of DR during peak periods can result in significant savings in terms of energy expenditure. In wholesale markets, spot energy prices during peak periods can skyrocket due to increased demand. Similarly, energy prices in vertically integrated, non-wholesale market systems can also increase during peak periods as less efficient units (i.e. with a higher heat rate) are utilized in order to meet the rising demand. As retail energy rates tend to not reflect the true cost of energy during peak periods, the expensive utilization of generation during these times is socialized among all customers. By reducing the need to purchase high-priced power, all customers in a system are positively impacted. The aforementioned Brattle report also identified energy savings on the order of $300 million per year from the same 5% reduction in the peak demand of the US as a whole. Figure 1 further illustrates the point that the avoided capacity costs far outweigh the avoided energy and avoided T&D costs.

Indeed, it is important to recognize the financial benefits participants in these programs receive, which are the sum of both the avoided energy costs (and demand charges) as well as the direct incentive payments for participation and successful performance.

2. Demand-Side Management and the Role of DR

2.1. Overview of Demand Side Management

Demand-side management (DSM) consists of a broad range of planning, implementing and monitoring of activities designed to encourage end-users to modify their levels and patterns of electricity consumption. DSM programs and initiatives are typically implemented to achieve two basic objectives: energy efficiency (EE) and load management. EE is primarily achieved through programs that reduce overall energy consumption of specific end-use devices and systems by promoting high-efficiency equipment use and building design. Conversely, load management programs are designed to achieve reductions in consumption primarily during times of peak demand, rather than on a permanent or ongoing basis. Load management programs can include permanent-load shifting and peak-shaving activities traditionally associated with demand response. With improvements in technology, load management programs are also increasingly dispatched on a level playing field with supply-side resources.

Figure 2 presents the total peak load reduction through DSM from 1998 to 2009 in the US. Peak load reduction through energy efficiency programs increased from 13,591 MW in 1998 to 19,766 MW in 2009, but decreased from 13,640 MW to 11,916 MW during the same period by load management programs.[7]

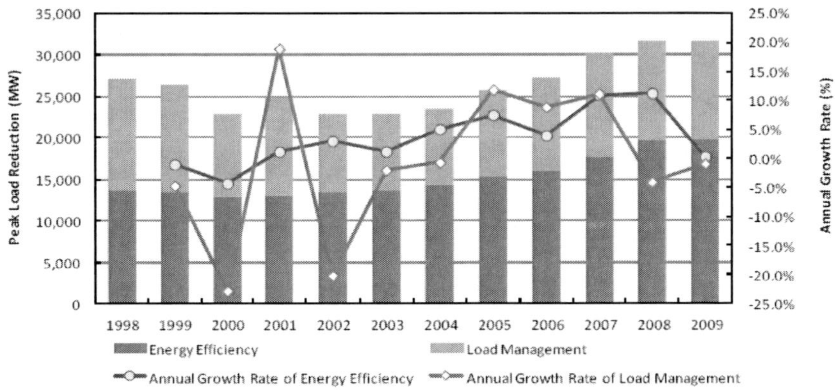

Figure 2. Peak Load Reductions from DSM Programs by Program Category.[8]

2.2. Role of DR in Demand-Side Management

Demand Response is normally included as part of utility DSM program or a potential DSM program solution which helps make the electric grid much more efficient and balanced by assisting the electric grid's commercial and industrial customers in reducing their electric peak demands, and/or shifting the time period when they use their electricity, and/or prioritizes the way they use electricity, and in return reduces their overall energy costs.

A key difference between DR and EE is the energy reductions for DR are time-dependent, whereas reductions for EE are not. Demand response programs yield reductions in demand at critical times, which typically corresponds to time of peak power demand, while EE programs yield permanent energy savings. However, the two programs have overlapping effects: EE can permanently reduce demand including those occurred during the peak time while demand response with well-targeted control strategies can also produce energy savings.[9]

Up to 2003, EE programs in the US contributed to more than 60% of actual peak load reduction;[10] however, its share dropped by almost 10% from 59.3% in 2004 to 49.9% in 2009. Meanwhile, contributions from load management had increased by about 12% from 37.6% to 50.1% during the same period (see Figure 3).

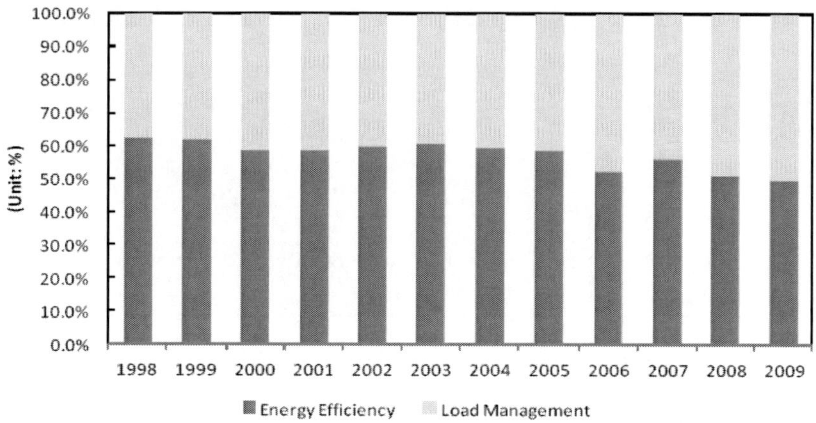

Figure 3. Share of Total Actual Peak Load Reduction by Program Category.[11]

2.3. Coordination of Demand Response and Energy Efficiency

Demand response and energy efficiency programs could be coordinated at the customer level at least in the following four ways:[12]

- **Offering combined programs**: Although separating energy efficiency and demand response programs are quite common, customers could be presented with both opportunities at the same time. Furthermore, technologies that are commonly used for DR – such as energy monitoring, building automation systems and load control equipment – can also be leveraged to help inform about opportunities that would lead to energy efficiency improvements.
- **Coordinating program marketing and education**: Program sponsors could package and promote demand response and energy efficiency in a closely coordinated way. Because the two programs are quite complicated to customers, program sponsors could help customers addresses both topics under a broad DSM and management theme.
- **Market-driven coordination services**: Effective coordination can be done not only by utilities and Independent System Operators (a.k.a. ISO), but also by the initiative of private firms that find a market

among customers who are interested in reducing their energy costs or receiving incentives.

- **Incorporating Building codes and appliance standards**: Building codes and appliance efficiency standards can incorporate demand response and energy efficiency functions into the design of buildings, infrastructure, and power-consuming appliances/equipments. Integrating those codes and standards can lead to significant reduction in the costs to customers of integrating demand response and energy efficiency strategies and measures.

3. Regulatory and Policy Frameworks That Promote Demand Response

3.1. Demand Response Enabling Policies

DR, at least in a basic form, has been around for decades. In the US, load management and interruptible/curtailable tariffs were first introduced in the early 1970s. The primary interest in load management was driven in part by the increasing penetration of air conditioning which resulted in needle peaks and reduced load factor. These programs were effectively limited to the largest industrial customers in a given system, and in many cases never used. Deployed before the advent of the internet or the load aggregator business model, these programs were very manual and typically featured slow response times. With such limited capabilities, interruptible programs served less as an alternative to generation investments, and more as a load management tool that could theoretically be used in emergencies – in reality though, they were more often than not a customer retention tool allowing utilities to offer discounted service rates to customers large enough to fund the installation of their own generation assets.

This base of demand response was then further spurred by two important developments: Within traditionally-regulated, vertically-integrated utilities, the advent of integrated resource planning in the late 1970s and 1980s made utilities increasingly aware of the system cost impacts of meeting peak loads, and load management began to be viewed as a reliability resource. The results of this perspective were first evident in the rise of Direct Load Control (DLC) programs that cycled residential air conditioning units during peak periods. Even more significant, in the mid 1990s, policymakers and utilities interested in facilitating the development of regional, competitive wholesale markets primarily based on re-design and re-structure markets.

3.1.1. Wholesale Market Access

United States

It is well-known to industry observers that the growth of the demand response industry in the United States can in many ways be traced to the opportunities in these wholesale power markets, particularly in the systems of ISO-New England[13] and the PJM Interconnection[14]. According to the most recent government statistics, more than 31 GW of demand response was active in the US RTO/ISO markets in 2011[15]. While the opportunities for DR in California that emerged after the state's energy crisis in 2001 certainly contributed to the growth of the industry as well, the scale of the opportunities (and the realization of them) in the aforementioned wholesale markets has proven to be a stronger influence on the growth of the industry in the United States.

The role of government policy in the establishment of these opportunities has been an essential driver to the growth of the DR industry in the US. The foundation of competitive power markets in the US can be traced to the Energy Policy Act of 1992 (EPAct) and Order 888 from the Federal Energy Regulatory Commission (FERC). EPAct began the process of electric industry deregulation and opened up the opportunity for independent power generators to participate in wholesale markets, which FERC Order 888 furthered by requiring fair access and market treatment to transmission systems. While the aforementioned legislation and Order were primarily focused on increasing competition among generators, the concepts laid the groundwork for demand response to enter wholesale markets when such resources could meet the same technical requirements as their supply-side counterparts. The Energy Policy Act of 2005 (EPACT) further codified that a key objective of US national energy policy was to eliminate unnecessary barriers to wholesale market demand response participation in energy, capacity, and ancillary services markets by customers and load aggregators,[16] at either the retail or wholesale level.[17]

While demand response began participating at scale in wholesale power markets in the early 2000s – particularly in emergency capacity programs – many market barriers remained. Fortunately, in October 2008, FERC issued Order 719, which focused on the operation of the country's wholesale electric markets. A major component of Order 719 was eliminating barriers to the participation of demand response in wholesale markets operated by wholesale market operators. Order 719 permitted load aggregators to bid demand response directly into organized markets, unless the relevant laws of the local electric retail regulatory authority prohibit such activity.

Demand response integration into US wholesale power markets was further bolstered with the March 2011 issuance of FERC Order 745. Order 745 requires that demand response resources are paid the Locational Marginal Price (LMP), or the wholesale market price for energy. By codifying the ability for DR to be compensated in the same fashion as generation resources for services provided to the energy markets, Order 745 advances the cause of equal treatment between generation and demand side resources.

In the US, DR is primarily seen in the wholesale capacity markets, most notably in the PJM Interconnection and ISO-New England. DR in these markets is procured in a competitive process that places demand side resources on equal footing with generation, creating an opportunity for cost-effective DR that can easily enter the market (should technical requirements be able to be met). In addition, in both markets, capacity DR is dispatched only during the very critical peak or emergency periods, making end-user participation relatively simple (compared to other markets to be profiled in this paper that are solely for balancing resources). Examples of both markets are provided below.

The PJM Interconnection

PJM (Pennsylvania-New Jersey-Maryland) Interconnection is a regional transmission organization (RTO) that coordinates the movement of wholesale electricity in all or parts of Delaware, Illinois, Indiana, Kentucky, Maryland, Michigan, New Jersey, North Carolina, Ohio, Pennsylvania, Tennessee, Virginia, West Virginia and the District of Columbia. PJM is the largest market in the US and allows DR to participate in all of its markets types – capacity, energy, and ancillary services. Today, more than 60 entities serve as Curtailment Service Providers (CSPs), or load aggregators, in the PJM system.

Like in most systems, the bulk of the DR in PJM participates in the capacity market. Capacity markets are particularly well-suited to peaking resources like DR which operate for relatively few hours a year and may have trouble accessing the proper price signals from an energy-only market. PJM's current capacity market, the Reliability Pricing Model (RPM), was instituted in 2007. In the RPM, those resources include not only generating stations, but also demand response actions and energy efficiency measures by consumers to reduce their demand for electricity. In this manner, demand side management is directly integrated into the wholesale capacity market structure.

Every year PJM conducts a Base Residual Auction (BRA) for delivery of capacity three years in the future. The BRA is held in May and the delivery year begins 3 years later on June 1st and ends on May 31st of the following year. In addition to the BRA, PJM conducts three Incremental Auctions (IA) that are held in advance of each corresponding delivery year. The purpose of the IA is to balance any changes in the load forecast and to allow suppliers of capacity resources to adjust their positions. In PJM's most recent Base Residual Auction in May 2011, the market procured 149,974 MW of capacity for the 2014/2015 delivery year. Of note, 14,118 MW of this capacity – or 9.4% of the total – came from demand response resources. Once cleared through the capacity market, these DR resources become participants in PJM's Emergency Load Response Program.

In PJM, qualifying DR resources can also participate in the wholesale energy market (both day-ahead and real-time) as well as various ancillary service markets (primarily, the synchronized reserve market). However, these markets are not the same drivers of DR growth that the capacity market is. The energy market does not feature capacity incentives, and therefore requires significantly more participation to garner the same financial opportunity. The Synchronized Reserve Market, on the other hand, requires full response within 10-minutes of a dispatch signal and generation-grade telemetry, limiting the pool of potential participants.

ISO New England (ISO-NE)

PJM's counterpart to the north, ISO New England, also operates a forward capacity market (FCM) in which DR can participate alongside generation, and which accounts for the majority of demand side participation within the New England system. Similar to PJM's BRA, the ISO-NE FCM also allows both dispatchable demand response and energy efficiency measures to participate in the market.

ISO-NE's use of demand response may be the clearest example of a resource designed specifically for reliability and/or emergency prevention purposes. While the ELRP in PJM is also designed for similar purposes, the trigger for usage in ISO-NE is even more defined. ISO-NE treats DR provided by curtailment and on-site generation as distinct resources, labeling the former Real Time Demand Response (RTDR) and the latter Real Time Emergency Generation (RTEG). RTDR may be called by ISO-NE only when the system reaches an emergency level known as Operating Procedure 4 Action 9 (OP4

Action 9). RTEG, on the other hand, cannot be dispatched until a further level of emergency has been reached, OP4 Action 12. For customers that utilize both load curtailment and on-site generation to provide DR capacity, they (or their DR provider), must be able to call those distinct loads separately in order to comply with ISO-NE requirements. Today, approximately 2,000 MW, or 8% of the resources in the capacity market, are dispatchable demand response. This figure grows to 3,400 MW, or 10% of the ISO-NE system, in 2014/15.

Demand response resources can also provide energy to the ISO-NE market through the Real-Time Price Response and Day-Ahead Load Response Programs. As with energy market participation in PJM, these programs are relatively unpopular compared to the capacity market, as they require much more frequent participation and have comparatively lower economic benefit. In both markets, sites that participate in the energy programs tend to be among the most flexible participants in the capacity markets who are looking for an additional economic opportunity, rather than the energy program serving as the sole method of DR participation in the market. While ISO-NE formerly had a pilot program testing the ability for DR to provide ancillary services – the Demand Response Reserve Pilot (DRRP) – it no longer has an active mechanism for DR to provide operating or spinning reserves. DR participation in these markets is now under active consideration, in part due to the aforementioned FERC Order 719.

United Kingdom

National Grid Short Term Operating Reserves (STOR) Market

Demand response resources also enjoy wholesale market access in the United Kingdom, albeit in a much more limited context. Market-based opportunities for demand-side resources in the UK are currently restricted to ancillary service markets, primarily the Short Term Operating Reserves Market. While others exist, the parameters result in low levels of participation and or a small addressable market.[18] DR cannot access the nation's wholesale energy market and unlike PJM and ISO-NE, there is no capacity market in the UK. That said, the government, spearheaded by the Department of Energy and Climate Change (DECC), is pushing forward legislation to launch one in the coming years and which will also allow for demand side participation.

STOR is essentially a supply and demand balancing service that meets the need of the grid as demand changes and as traditional power plants come online and ramp up and down, similar in many ways to the Sycnhronized Reserve Market (SRM) in the PJM Interconnection. While not a capacity market per se, cleared resource receive an availability payment for each hour they are in the market and available to be dispatched. Utilization (energy) payments are also given for the actual load reduction provided. Both features are also present in the aforementioned SRM. As a balancing market, STOR is called much more frequently than the capacity programs in PJM and ISO-NE which are used primarily to address emergency conditions. STOR participants, on average, must be prepared to respond to a dispatch every week. Such frequent participation requires the employment of different curtailment strategies than those that are found in capacity programs designed to shave consumption only during infrequent, peak periods.

Australia

Independent Market Operator – Wholesale Electricity Market (WEM)

Another successful example of DR participation in wholesale markets can be found in the South West Interconnected System of Western Australia, run by the Independent Market Operator (IMO). There are two wholesale markets in Australia; the Whosesale Electricity Market (WEM) in Western Australia, and the National Electricity Market (NEM) in the eastern states (except for the Northern Territory). The NEM is an energy-only market and has very low levels of DR participation for the reasons mentioned earlier, whereas the WEM is a capacity market similar in many ways to ISO-NE and PJM, and with a significant penetration of DR. In the most recent Reserve Capacity Cycle, more than 8% of the capacity procured came from demand-side resources.[19]

The IMO-administered WEM procures system resources through the Reserve Capacity Mechanism, in which capacity can be traded bilaterally to the IMO directly, or to retailers. Unlike in PJM and ISO-NE where capacity prices are the result of competitive offers, the IMO sets a price for all capacity based on the avoided cost of a marginal new peaking unit, specifically a 160 MW open cycle gas turbine. Auctions are only triggered if the bilateral trading mechanism secures insufficient capacity.[20] As in the previously discussed capacity markets of the US, DR and generation receives the same exact market payment.

As with PJM and ISO-NE, DR is assumed to have different levels of dispatch capability than traditional supply-side resources. The RCM has 4 Availability Classes; Generation must all list itself as Class 1, or available for more than 96 hours a year; DR meanwhile can offer at between 24-96 hours of dispatch.

One important distinction about DR in WA is that unlike PJM and ISO-NE; DR in the WEM can be dispatched when it is deemed to be economic and is not dependent on emergency conditions. The system operator is required to first utilize the plants of the former state-owned generation company, but afterwards all dispatch is determined by the market energy price offered by the resource.

Unlike its counterparts in the US, DR in WA is limited to participation in the wholesale capacity market. While a wholesale energy market also exists in WA, the STEM, DR resources do not have access to the market. Meanwhile, a competitive balancing market is only just now being designed, and access for DR is expected once the market is fully operational in the coming years.

Capacity-based Programs in Energy-only Markets

Some wholesale market operators have taken a slightly different approach, creating DR-specific opportunities outside of the standard wholesale markets themselves. For the most part, these are energy-only markets, where the underlying structure is not as conducive to peaking resources like demand response, which operate for relatively few hours per year.

In the Canadian Province of Ontario, the Independent Electricity System Operator (IESO) and the Ontario Power Authority (OPA) launched a large scale DR program (DR3) in 2007 to provide additional capacity to the market due to planned retirement of coal-fired power plants in the province. DR3's inclusion of a capacity payment represented a departure from previous DR programs in Ontario that failed to gain traction, primarily due to incentives being limited to energy payments.

Another example is the Emergency Interruptible Load Service (EILS) program in the Texas market of ERCOT. EILS is essentially a standalone markets that exists alongside the wholesale markets open to generation resources in ERCOT. EILS is designed to provide reserve capacity to the energy-only ERCOT market, and is procured during four separate markets spaced evenly throughout the year. Unlike DR3, pricing in EILS is the result of offers made by DR providers, similar to how the capacity markets in PJM and ISO-NE work. A further similarity with PJM is that ERCOT allows EILS participating loads to provide operating reserves in the ERCOT ancillary service markets and receive the same payment as generation resources, similar to the SRM.

3.1.2. Bilateral Programs with Vertically Integrated Utilities and Network Operators

Access to existing wholesale markets are just one mechanism for creating and leveraging demand response resources. In recent years, much growth in the industry has been found in bilateral programs with vertically integrated utilities in traditionally regulated environments, and with network (T&D) operators located within a liberalized market structure. These bilateral programs are most often used as a way to avoid or defer investments in generation and/or T&D infrastructure, and tend to look similar in structure to a power purchase agreement (PPA) that a utility might sign with an independent power producer. These utility programs are likely better proxies for how the implementation of next-generation demand response could manifest itself in China, given the lack of a wholesale market.

There are a number of enabling policies that have encouraged the development of bilateral DR programs throughout North America, the UK and Australia. These policies include:

- Cost recovery and DSM funds
- Loading orders and similar regulations
- Peak demand mandates and energy efficiency portfolio standards

Cost Recovery and DSM Funds

Whether in the US, the UK, or Australia, vertically integrated utilities and distribution network operators are regulated monopolies whose revenues are dependent on government policy and regulation. As such, it is essential to understand the regulatory environments in which these utilities operate in order to understand how regulatory policies have both contributed to, and hindered, the growth of demand response.

Perhaps the most basic and essential enabling policy is a cost-recovery mechanisms. Under a cost-recovery mechanism, a utility can recover prudently-incurred costs of DR and EE investments on a dollar-for-dollar basis, typically through a rider or customer surcharge. Cost recovery is designed to make a utility whole on its DR and EE investments. However, there are challenges with this approach. First, cost recovery alone will not address the lost margin revenue the utility will face due to reduced energy sales from DR and EE programs. Second, cost recovery does not factor in opportunity costs: DR and EE investments displace supply-side investments for which the utility can earn a profit. Given these opportunity costs, absent a statutory or regulatory mandate, program cost recovery alone will generally not attract utility interest in DR and EE programs.

However, in some jurisdictions, utilities are authorized to recover additional costs associated with the lost revenue due to the energy efficiency measures. There are also provisions for earning a fair rate of return on the DSM investment, typically at levels that are equivalent to allowable returns on power generation assets.

Loading Orders and Similar Regulations

Loading orders are governmental proclamations that define the priority order in which resources are to be developed. To underscore the importance of energy efficiency and demand response in California's future energy picture, the state government developed the Energy Action Plan established a "loading order" of preferred resources, placing energy efficiency and demand response as the state's highest-priority procurement resource, and set aggressive long-term goals for energy efficiency and demand response resources. In addition, energy efficiency and demand response strategies were implemented to address greenhouse gas emission reduction targets specified by AB32, a law adopted in California to create regulatory policy mechanisms to combat global warming. As a result of these policies, California's energy efficiency and demand response efforts have proven to be very successful. California leads the nation in term of energy saved. The state invests nearly $3 billion per year in energy efficiency and demand response programs that target electricity and natural gas customers to install high efficiency equipment, take measures to reduce their peak demands, and establish time-sensitive price structures that are more in line with the actual cost of providing the electricity. Resources such as renewable generation, distributed generation, and traditional generation are considered as the second and third priorities, respectively in the loading order, and should only be considered once all energy efficiency and demand response resources are exhausted.

In Massachusetts, a law known as the Green Communities Act was passed in 2008 and implemented shortly thereafter. The law requires the state's utilities to procure all available energy efficiency resources that cost less than traditional energy sources do. The law in effect prioritizes energy efficiency as being at the top of the loading order, ahead of renewable energy, and more traditional forms of generation. Among the major provisions is a requirement for utilities to invest in energy efficiency when it is less expensive than buying power. Previously companies purchased more power when demand increased. The effect of the law is that the state is seeing significant investments in energy efficiency, leading toward the ultimate goal of reducing the state's use of fossil fuels in buildings by 10% and overall greenhouse gas emissions by 20% in the year 2020.

Peak Demand Mandates, Energy Efficiency Portfolio Standards

Peak demand mandates and energy efficiency portfolio standards have recently emerged as another mechanism to encourage DR outside of market-based opportunities. Perhaps most well known is a mandate in the state of Pennsylvania, the so-called Act 129 legislation, signed into law in October 2008, which requires all electric distribution companies to achieve peak demand reduction targets of 4.5% and energy efficiency reductions of 4% by 2015. While the legislation does not expressly encourage DR over other types of peak reduction such as energy efficiency and or solar PV, Pennsylvania utilities appear to have determined C&I DR was the most cost effective way to reach compliance and several large deals with aggregators have already been publicly announced.

Other states with peak demand mandates that are similar to Pennsylvania include New York, Colorado, Michigan and Ohio. In New York, the Public Service Commission established an Energy Efficiency Portfolio Standard (EEPS) which ordered the state's utilities to achieve a 15% reduction in forecast electricity usage by the year 2015. The state's utilities are implementing aggressive EE and DR programs in order to meet that goal, which specifies that each of the state's utilities realize specific MWh and peak MW reduction amounts by 2015. In Colorado, the Climate Action Plan (CAP) sets carbon reduction goals for the state and proclaims that energy efficiency programs are the most important responses to the carbon-reduction challenge. In response, the Colorado Public Utilities Commission has ordered the state's utilities to implement EE and DR programs to meet that goal. Michigan and Ohio have similar statutory mandates to lower energy usage and peak demand.

Parity of Treatment

Traditional utility regulation favors supply-side resources over DR and EE resources. First, utilities earn a rate of return on investments in generation, transmission and distribution infrastructure. The absence of a parallel incentive for DR and EE investments creates a bias against demand-side resources. This has been described in the economic literature as the "Averch-Johnson Effect." That is, where a firm's profits are linked to its capital investment, as is the case with utilities under traditional regulatory structures, there is an embedded incentive for the firm to increase its capital outlay in a manner that does not necessarily maximize producer and consumer surplus. Stated another way, traditional regulatory frameworks create a disincentive for utilities to meet resource needs using approaches that are less capital intensive. Thus, faced with otherwise equivalent alternatives of building a power plant that contributes to profitability or making investments in DR and EE that allow for cost-recovery only, a utility would generally prefer to build a power plant (or T&D).

The government of the United Kingdom recently recognized and addressed this very challenge. In the 2010-2015 Distribution Price Control Review 5 (DPCR5), Ofgem – the national electricity and gas regulator – instituted the so-called "Equalisation Incentive" which establishes parity in the treatment of capital and operating expenditures by distribution utilities. Thus, any utility acting in its own rational economic interest will clearly pursue the most cost-effective way to meet network needs and reliability requirements, whether that is through traditional investments in infrastructure or through non-network alternatives like DSR. As a result of this new regulation, one local distribution network operator – Electricity North West – has already deployed a commercial scale DR program in which an aggregator is deploying DR on specific circuits in order to defer investments in substations. Other distribution network operators, such as UK Power Networks, are also conducting pilot projects using DR for distribution relief as they hope to prepare themselves to launch commercial-scale programs under this new regulatory framework.

Example Utility DR Programs

There are several examples of bilateral DR programs. In California, Pacific Gas and Electric (PG&E) implements the Aggregator Managed Portfolio (AMP) program. AMP is a non-tariff program that consists of bilateral contracts with aggregators to provide PG&E with price-responsive demand response. The program can be called at PG&E's discretion. Each aggregator is responsible for designing and implementing their own demand response program, including customer acquisition, marketing, sales, retention, support, event notification and payments. To participate, customers must enroll through a load aggregator. The customer in turn authorizes the aggregator to act on their behalf with respect to all aspects of AMP, including receipt of notification of an event, receipt of incentive payments and/or penalties. Southern California Edison (SCE) operates the Demand Response Contracts (DRC) program. SCE has contracted with several aggregator companies to provide SCE with price-responsive and/or demand response events that SCE may call at its discretion. Each aggregator designs their own programs, and offers demand response program structures and options that may not be directly available through SCE. Customers may select an aggregator with services that best meet their business needs.

More common are arrangements where a utility contracts with a single DR load aggregator for a program in their territory (or a single provider per customer class). For example, EnerNOC, a Boston-based load aggregator has a program in place with the Tennessee Valley Authority (TVA) in the southeastern US, the largest public power company in the country. TVA procured a long-term, 560 MW resource from EnerNOC which it is required to deliver in line with contract requirements over the 10- year contract length. There are many other load aggregator companies operating in the various electricity markets throughout North America. As with the aforementioned DR programs in California, the load aggregator is responsible for all roles from customer acquisition through resource dispatch and settlement.

As with similar DR programs, TVA has purchased a guaranteed firm resource. In addition to identifying and enabling DR capacity in line with contract milestones, the load aggregator must also meet performance standards when dispatched by TVA. Should the load aggregator fail to do either, financial penalties against the aggregator may be assessed. In this manner, TVA can depend on its DR-based "virtual power plant" in the same way its system planners and operators can trust a traditional generation resource. Figure 4 provides a summary of the TVA bi-lateral program parameters.

Program Size	Up to 560 MW
Advanced Notification	30 minutes
Dispatch Trigger	TVA's discretion
Availability Window	April – October: 12:00-20:00, Mon-Fri November – March: 5:00-13:00, Mon-Fri
Maximum Cumulative Dispatches	40 hours per annum
Term Length	10 years

Figure 4. TVA Bi-lateral DR Program Parameters.

Other vertically-integrated utilities in the US that have implemented similar programs include: Arizona Public Service, Idaho Power, NV Energy, Public Service Company of New Mexico, Puget Sound Energy, Salt River Project, San Diego Gas & Electric, Tampa Electric, Tucson Electric Power, and Xcel Energy. Per the aforementioned "equalisation incentive" now in effect in the UK, distribution network operators (DNOs) in the country are also now deploying demand response programs to defer or avoid investments in network infrastructure. Electricity North West (ENW), one of the 14 regulated DNOs in the UK with a network that includes the Greater Manchester and Cumbria areas, has recently launched a DR program along with a third-party load aggregator. Under this program, DR is deployed within specified circuits in the network, allowing demand to be controlled on a geographically-targeted basis that will prevent the need to upgrade the substations on those portions of the network. The program, announced in May 2011, is set to last for five years. Such 'network support' contracts are also commonly found in Australia, particularly in New South Wales.

The same Distribution Price Control Review that launched the equalisation incentive, also included funds from Ofgem – the UK electric regulator – for Low Carbon Network (LCN) projects that will pilot new technologies and facilitate the development of an environmentally-friendly electricity system in the country. Many DNOs throughout the UK have successfully applied for LCN funding to pilot the use of DR in their networks, including UK Power Networks (UKPN) and Northern Powergrid (formerly CE Electric). ENW has also recently been awarded LCN funding from Ofgem to pilot the use of DR in new ways within their system that, if successful, would reduce the amount of network capacity DNOs would need to have in order to comply with reliability standards.

3.2. Encouraging End-User Participation: The Role of Incentives

The U.S. Department of Energy classifies demand response into two categories, i.e. price-based demand response and incentive-based demand response.[21] Each category has its own subcategories. Pricing mechanisms vary on each subcategory as shown in Table 1.

- *Price-based demand response* refers to changes in usage by customers in response to changes in the prices they pay and include real-time pricing, critical-peak pricing, and time-of-use rates. If the price differentials between hours or time periods are significant, customers can respond to the price structure with significant changes in energy use, reducing their electricity bills if they adjust the timing of their electricity usage to take advantage of lower-priced periods and/or avoid consuming when prices are higher. Customers' load use modifications are entirely voluntary (Table 1)
- *Incentive-based demand response* programs are established by utilities, load-serving entities, or a regional grid operator. These programs give customers load-reduction incentives that are separate from, or additional to, their retail electricity rate, which may be fixed (based on average costs) or time-varying. The load reductions are needed and requested

either when the grid operator thinks reliability conditions are compromised or when prices are too high. Most demand response programs specify a method for establishing customers' baseline energy consumption level, so observers can measure and verify the magnitude of their load response. Some demand response programs penalize customers that enroll but fail to respond or fulfill their contractual commitments when events are declared (Table 1).

Table 1. Demand Response Options and Related Pricing Mechanisms

Price-Based (Voluntary)	Incentive-Based (Contractually Mandatory)
• *Time-of-use (TOU)*: a rate with different unit price for usage during different blocks of time, usually defined for a 24 hour day. IOU rates reflect the average cost of generating and delivering power during those time periods.	• *Direct load control*: a program by which the program operator remotely shuts down or cycles a customer's electrical equipment (e.g., air conditioner, water heater) on short notice. Direct load control programs are primary offered to residential or small commercial customers.
• *Real-time pricing (RTP)*: a rate in which the price for electricity typically fluctuates hourly reflecting changes in the wholesale price of electricity. Customers are typically notified of RIP prices on a day-ahead or hour-ahead basis.	• *Interruptible/curtailable (I/C) service*: curtailment options integrated into retail tariffs that provide a rate discount or bill credit for agreeing to reduce load during system contingencies. Penalties maybe assessed for failure to curtail. Interruptible programs have traditionally been offered only to the largest industrial (or commercial) customers.
• *Critical Peak Pricing (CPP)*: CPP rates are a hybrid of the IOU and RIP design. Ihe basic rate structure is IOU. However, provision is make for replacing the normal peak price with a much higher CPP event price under specified trigger conditions (e.g., when system reliability is compromised or supply prices are very high).	• *Demand Bidding/Buyback Program*: customers offer bids to curtail based on wholesale electricity market prices or an equivalent. Mainly offered to large customers (e.g., one megawatt [MW] and over).
	• *Emergency Demand Response Programs*: programs that provide incentive payments to customers for load reductions during periods when reserve shortfall arise. (e.g. ERCOI EILS)

Price-Based (Voluntary)	Incentive-Based (Contractually Mandatory)
	• *Capacity Market Programs*: customers offer load curtailments as system capacity to replace conventional generation or delivery resources. Customers typically receive day-of notice of events. Incentives usually consist of up-front reservation payments, and face penalties for failure to curtail when called upon to do so. (e.g. PJM ELRP, IMO WA)
	• *Ancillary Services Market Program*: customers bid load curtailments in ISO/RIO markets as operating reserves. If their bids are accepted, they paid the market price for committing to be on standby. If their load curtailments are needed, they are called by the ISO/RIO, and may be paid the spot market energy price. (e.g. PJM SRM, UK SIOR)

Source: DOE (2006), p.12.

In addition to federal regulation as described in Section 3.1 and economic benefits described in Section 3.2, numbers of the U.S. utilities have taken action to expand their retail demand response programs. One incentive factor for many of them has been concern about peak load growth and rising energy prices.[22]

3.2.1. Lack of Sufficient Incentives from Standard and TOU Pricing: Experience with Interruptible Tariffs

Many utilities have offered a variety of traditional DR programs for many years. These legacy programs are typically referred to as load management programs. There are three types of legacy load management programs: direct load control (DLC), time-of-use (TOU) rates, and interruptible contracts. Each of these programs use some form of incentive to encourage customers to participate. However, the amount of the incentives or the nature of the incentives has not been sufficient to bring about meaningful levels of demand reductions.

DLC programs allow the utility to directly control customer end-uses during certain periods when the electrical system is under strain. The customer end-uses are directly controlled by the utility and when events are called, those loads are either shut down, cycled on and off, or moved to a lower consumption periods. Residential DLC programs often target air conditioners or electrical water heaters. Nonresidential DLC programs include air conditioner systems, lighting and in some regions irrigation control. There are a number of challenges with DLC programs. First, customers tend to become frustrated with effects of the service interruptions and oftentimes will leave the program if they are called too frequently. Second, the incentives offered by the utilities have been insufficient to encourage their sustained participation.

TOU rates are tariff schedules that are typically offered to residential and small business customers on a voluntary basis and are mandatory for the largest commercial and industrial customers. The TOU rates are structured to charge lower rates during a utility's off-peak and partial-peak periods and higher rates during seasonal and daily peak demand periods. By charging more during the peak period, when incremental costs are highest, TOU rates send accurate marginal-cost price signals to customers. TOU rates encourage customers to shift energy use away from peak periods to partial-peak or off-peak periods and enable customers to lower their electricity bills. There are two common challenges with TOU rates. First, the utilities have often set the TOU peak periods to be for long periods at a time, thus limiting customers' abilities to shift their loads to the lower price off-peak periods. Second, the TOU rate programs tend to be static in nature in that the peak and off-peak prices do not change regardless of system conditions and the true costs required to deliver electricity to customers. Because of the static nature of the TOU rates, they cannot be counted on for meeting the peak demand needs of the utility.

In addition, the utilities often design these tariffs to be revenue neutral. That is, the price differentials between on-peak and off-peak are intended to not change the utility's overall revenue. This goal oftentimes is inconsistent with a goal of maximizing customer participation in order to have meaningful peak demand reductions as a result of the TOU tariff.

Interruptible tariffs are contractual arrangements set up between the utility and large non-residential customers. Customers agree to reduce their electrical consumption to a pre-specified level, or by a pre-specified amount, during system reliability problems in return for an incentive payment or a similar rate discount. Customers are given the incentive regardless of whether reliability events are called. In the past, these programs were developed mostly for customer retention as the utilities assured customers that reliability events were so rare and would never be called. However, as reliability problems are becoming more acute, utilities are calling more interruptible events. As a result, many customers are opting to negotiate an exit to their contractual obligations for these programs as they cannot tolerate the volume interruptions to their businesses.

3.2.2. Cost and Risks–How Load Aggregators have Removed Traditional Barriers to DR Participation

Complicated tariff structures and insufficient incentives are just a few of the challenges utilities face when trying to garner customer interest in traditional, non-aggregator-based DR programs. Equally important are the costs and risks customers must bear in order to participate.

While the costs for metering and load control equipment may not always be borne by the customer in these situations, the exposure to performance penalties remains essentially a constant. Without an aggregator to guarantee the load response, utilities have no choice but to penalize customers if they don't fully comply with a dispatch in order to ensure proper response. However, C&I loads are inherently volatile – and customers may not always be able to participate – and consequently customers may need to be willing to face a strong likelihood of penalties if they seek to participate.

Using load aggregators is one proven approach to removing many of these traditional barriers to DR participation. It is typical that the load aggregator pays all costs for the installation of metering and load control equipment, making participation for the customer a no-cost proposition. More importantly, because load aggregators are measured on the total load reduction their entire portfolio of sites provides, and not on a site-by-site basis, they are able to pool resources in a way that ensures that contract performance requirements can be met. In the event that performance penalties are assessed on the aggregator, many will still refrain from passing these onto the customer. Figure 5 illustrates this concept.

3.2.3. Avoid Energy Costs vs. Resource Payments

As is evident from the relative lack of uptake in energy-based demand response opportunities in wholesale markets (where prices are higher and more volatile than what customers face at a retail level), the economic benefit of avoided energy costs alone is likely to be an insufficient driver of customer participation. While dynamic pricing may impact this trend in some ways (as discussed in the next section), currently it is the ability to receive resource payments from DR participation that are driving customer involvement. In both wholesale market and bilateral programs, aggregators – or very large customers that qualify for direct participation – receive a payment from the entity purchasing the DR resources, either the system operator or the utility.

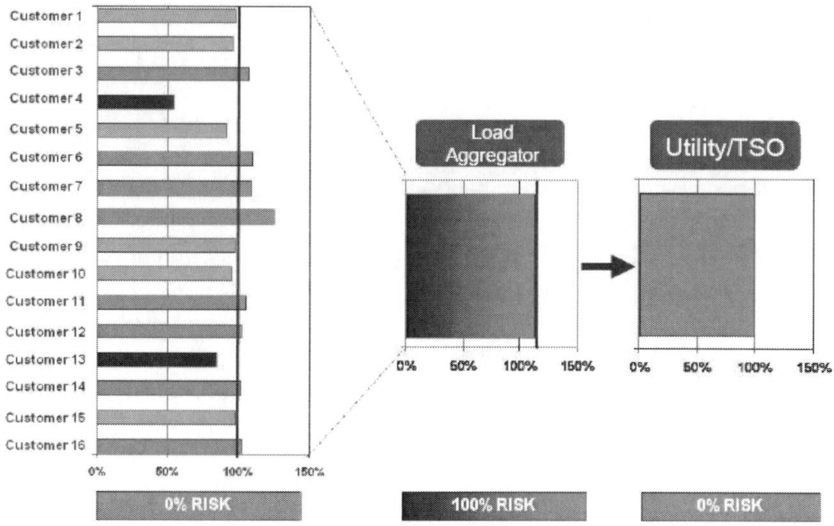

Figure 5. Aggregated Performance and Risk-Shielding.

Whether determined through market pressures or a utility decision, these payments are almost always based on the avoided costs of providing the same functional service through a traditional supply-side resource. Aggregators then use a portion of this payment stream to cover their costs of customer acquisition, site enablement, dispatch and settlement; the remainder is used to pay the customer an incentive payment for their participation. These payments tend to exist in the same form as those the aggregator receives, namely energy and capacity. In wholesale markets where very large customers can participate directly, the customer would individually receive the full payment stream, but would be responsible for adhering to technical requirements and managing performance risk. For these reasons, many large customers continue to work with aggregators even when the market requirements don't make it a necessity.

3.2.4. Emerging Trends: Dynamic Pricing

Dynamic pricing refers to a category of rates that offer customers time-varying electricity prices on a day-ahead or real-time basis. Prices are higher during peak periods to reflect higher-than-average cost of providing electricity during those times, and lower during off-peak periods, when it is cheaper to provide electricity. Dynamic pricing incentivizes customers to lower their usage during peak times, particularly during the most critical hours of the year when peak demands spike and the cost of acquiring electricity tends to be the highest. Dynamic pricing can take many forms. The most sophisticated form of dynamic pricing is real-time pricing (RTP). RTP programs are where prices are set by the utility in near real-time to match the market conditions for available power. Customers must be able to accommodate whatever price is given, which means that they take a significant risk that if prices spike they will either accept the higher price or be capable to rapidly reduce their consumption levels to avoid the high prices. Because of the complexities of RTP programs, most of the examples are in the pilot stages. Once sophisticated metering infrastructures are put into place and customers have the necessary building automation systems, it is likely that there will be more RTP programs coming on line in the future.

Critical peak pricing (CPP) is a less complex form of dynamic pricing. CPP programs are designed such that the prices for the top 60 to 100 hours are defined ahead of time, but the actual times in which these prices are in effect is not known until the day before the DR event or sometimes on the same day as the DR event. The price differentials are intended to be quite steep (oftentimes set at three to five-times the peak price) to encourage the customer to reduce or shift their loads during the critical peak times. CPP programs are offered to all customer types from residential to large commercial and industrial. A variant of CPP is peak time rebates (PTR). In PTR programs, a standard rate is applied during all hours but customers can earn a rebate if they reduce their consumption during the critical peak hours. PTR programs are most applicable to residential customers.

3.3. Summary/Comparative Analysis of Policy and Regulatory Frameworks

The global survey of demand response programs in this report illustrates a variety of regulatory constructs under which DR can thrive. Fundamentally, all of these regulations and policies in one way or another attempt to change the traditional paradigm that has historically lead to investments in additional supply-side infrastructure rather than load management.

Clearly, one method that has been incredibly successful in this regard is the wholesale capacity market. By removing the type of resource from the decision-making equation altogether and rather basing procurement decision on price alone, any resource that can meet the necessary market requirements can be purchased. With more 8-10%, or more, of system capacity met by DR in the PJM Interconnection, ISO New England, and the Western Australia Independent

Market Operator, these markets have shown a clear ability to drive significant penetration of demand response.

Outside of the established liberalized markets where DR is present, there tend to be more fragmented regulatory efforts to mitigate – but not eliminate – the disparity in incentives between supply-side and demand-side investments by utilities. In fact, one could argue that these multitude of policies and initiatives are required because in most areas, the underlying financial drivers that encourage a supply- side-focused perspective have not been modified: utility revenue is still tied to the amount of kWh sold, and the amount of capital they invest in generation and/or network infrastructure.

In many regions, utilities are only allowed to recover their DSM expenditures, but cannot earn a rate of return in the same manner as they would for supply-side investments. Because of this unequal treatment, some jurisdictions require their utilities to first pursue DSM programs before they can build generation assets to ensure solutions that may be cost-effective, but not financially beneficial, are considered. In other areas, utilities are mandated to reduce the peak demands (and energy consumption) or face penalties – such as in Pennsylvania – where there is no financial driver for the utility to do anything other than build more and more infrastructure.

It is within this environment that the UK's "equalisation incentive," is significant as it demonstrates a way to create true parity of treatment outside of a wholesale market context. While wholesale generation is competitive in the UK, distribution network operation is not – they are regulated monopolies in the same manner as vertically-integrated utilities in traditionally-regulated markets. Moreover, in traditionally regulated areas, such a mechanism could be applied to all investments so that generation (or alternatives to it) were also covered. In many ways, it is the concept of the "equalisation incentive" that is most important, and not its exact methodology. A multitude of regulatory mechanisms could likely be developed that would result in equal financial treatment between supply-side and demand-side investments, and it is important to not prescribe specific methodologies that may be better suited for one system than another.

This global survey demonstrates that good program designs are crucial to the success of demand response, perhaps more so than the existence of a formal market structures. Regardless of how DR programs or opportunities are engendered, programs must have the essential elements outlined in this paper in order to be sustainable, whether they are in liberalized markets or operated by vertically-integrated utilities.

In the wholesale capacity markets profiled in this paper, the programs found in the investor-owned utilities of California, as well as the program for the public utility TVA, clear similarities are evident. All such programs and markets are capacity-based, in which demand response resources are paid an ongoing payment for being available to provide capacity. In addition, all these examples are mainly targeted at the infrequent, yet expensive, top peak hours of the year. While there is indeed the ability for DR to provide more frequent response, such as in ancillary service markets, these general peak-shaving or emergency-prevention programs are suitable for the widest number of participants and can therefore lead to the highest levels of customer penetration.

Lastly, the inclusion of demand response load aggregators is another key recipe for success. In wholesale markets, often only the largest industrial customers can participate directly and aggregators are a mechanism for small and medium sized C&I customers to participate as well. Yet, even in such conditions, it is common for customers that could otherwise directly access the market do so instead through aggregators for the risk-mitigation benefits discussed in this paper. And in both the wholesale markets and among the regulated utility environments that are indeed more similar to the landscape in China, we see aggregators play two other key roles that contribute to the success of DR. From the utility or system operator perspective is the ability to provide guaranteed capacity. Once reliability can be ensured, system planners and operators are subsequently able to depend on the DR resource and reduce the usage of, or construction of, supply-side infrastructure. Put another way, without these guarantees, there would be limited ability for investments in demand response that lead to opportunities for participation among end-users. Equally if not more important is the behind-the-meter expertise that aggregators offer. With specialized staff and technology able to implement repeatable curtailment strategies that do not negatively impact commercial business operations, aggregators can both identify and leverage more capacity, and achieve higher levels of customer participation.

4. Enabling Technology Solutions for Demand Response

Demand Response enabling technology solutions are dependent on the level of automation a particular facility participating in DR program is capable of. Understanding the functional capabilities of building control systems, including the underlying technologies and software capabilities as installed, is essential to identify and quantify a specific facility's potential to participate in Automated Demand Response (Auto-DR) and to maximize load reduction savings without affecting day-to-day business or operations. The three key ways a DR program can be implemented are:

1. **Manual DR:** This involves manually turning off or changing comfort set points, lights, or processes or each equipment, switch, or controller.
2. **Semi-Automated DR:** This involves automation of HVAC or one or several processes or systems within a facility using Energy Management Control Systems (EMCS) or centralized control system, with the remainder of the facility under manual operations.
3. **Fully Automated DR:** This involves automation of an entire facility, with integration of end use loads into an EMCS and centrally managed with no human intervention.

Regardless of the type, technology plays an important role in the reliable operation of demand response.

4.1. Metering and Control Solutions

Metering

Granular meter data is essential to the successful operation of a demand response program. First and foremost, it is the foundation of accurate measurement and verification (M&V), which is necessary for both proper measurement of the performance of the DR resource as well as financial settlement. Ensuring that DR performs as expected requires real-time data, so that the actual consumption of participating facilities can be compared to accurate forecasts of what their consumption would have been should a dispatch not have occurred. Furthermore, real-time metering and data presentment allows for performance monitoring during a dispatch. For aggregators, this enables them to ensure their entire portfolio is cumulatively delivering the load reduction required, and if not, allows them to utilize other resources to provide the proper level of curtailment. From an end-user perspective, particularly among larger sites responding with some level of manual action, real-time data also allows for them to ensure they have taken the proper steps necessary to comply with their intended response.

That said, it is important to differentiate between real-time meter data for DR and typically advanced metering infrastructure (AMI). First, while AMI can facilitate DR it is by no means a prerequisite for successful deployments.

In fact, because AMI deployments are in their early stages, and often focused on the residential customer classes, most of the technology-enabled DR present today utilizes real-time meter data by the installation of additional technology. This typically includes directly accessing a utility meter through analog pulse or digital serial outputs, as well as metering/sub-metering specific loads such as a generator, and then transferring this information back to the DR aggregator using existing broadband and wireless infrastructure.

Even when and where AMI is present, it may be insufficient. Most smart meters and their supporting infrastructure are designed primarily with automated meter reading, and not DR in mind. As such, it is common for these new "smart" meters to read data every half-hour or hour, and then backhaul the consumption data once a day. Such infrequent and delayed measurements, while appropriate for AMR purposes, do not provide the needed functionality for DR aggregators whom need to ensure delivery standards are met in real time. In this manner, the installation of additional or specialized metering equipment is likely required even where AMI is present.

Load Control

Load control hardware is another essential component of modern-day, technology-enabled DR deployments and is often part of the same advanced metering kit that is installed on customer premises. Many customer types require some level of automation in order to be able to respond to a dispatch signal. A grocery store, for example, will typically not have an energy or facilities manager on staff able to initiate curtailment measures. Even if personnel was present, without automation, they would likely be unable to manually enact common strategies for this customer segment, including HVAC cycling, partial lighting curtailment, and anti-sweat heater (condensation) control. In other situations, it is the program requirements that require load control in order to comply with the response time. Ancillary service programs, and some bilateral utility programs, can have response times of ten minutes, or less. In fact, frequency responsive DR programs can have even shorter response times. For example, the Alberta Energy System Operator (AESO) just launched a DR program with a 200-millisecond response time. With such requirements, automation and load control is an absolute necessity. Yet even in traditional peak management programs, remote load control is increasingly being utilized for customer convenience and enhanced resource reliability.

The aforementioned metering/gateway devices installed are often the foundation for initiating load control as they feature two-way communication.

Such devices may toggle relays attached to specific circuits, send scripts to Building Energy Management Systems (BEMS) to begin pre-defined curtailment actions, or attach directly to industrial control equipment.

Dispatch, Monitoring and Management

In order to successfully leverage the metering and load control hardware described above, DR providers commonly deploy Network Operation Centers (NOCs) to utilize the aforementioned foundation technologies. It is from these NOCs that load aggregators can initiate automatic dispatch notifications to participating customers, remotely control customer loads and generation, monitor performance in order to ensure performance compliance, and coordinate technicians in the field.

Centralized control centers also allow DR to comply with telemetry requirements in a cost-effective way. Some grid operators require resource in some of their markets (e.g. PJM Synchronized Reserves, National Grid STOR) to be directly integrated into their respective control rooms with remote terminal units, or other similar equipment. Such generation-grade hardware is expensive, and would be cost-prohibitive to deploy at individual customer sites.

4.2. Auto-DR and OpenADR (with the AMI linkage)

Increasingly, Auto-DR activities in California and in pilots across the U.S. are carried out through Open communication technologies, namely the Open ADR technology developed by LBNL. Since 2010, OpenADR is being formally standardized within standard organizations and it is selected by the U.S. national Smart Grid activity coordinated by the National Institute of Standards and Technology (NIST) as the only standard to communicate price and reliability-based information.[23]

In the *Open Automated Demand Response Communications Specification (Version 1.0),*[24] OpenADR is defined as "a communications data model designed to facilitate sending and receiving DR signals from a utility or independent system operator to electric customers. The intention of the data model is to interact with building and industrial control systems that are pre-programmed to take action based on a DR signal, enabling a demand response event to be fully automated, with no manual intervention. The OpenADR specification is a highly flexible infrastructure design to facilitate common information exchange between a utility or regional transmission organization (RTO)/Independent System Operator (ISO) and their end-use participants. The concept of an open specification is intended to allow anyone to implement the signaling systems, providing the automation server or the automation clients."[25]

The specification also describes the scope of the OpenADR standard: "The Open Automated Demand Response Communications Specification defines the interface to the functions and features of a Demand Response Automation Server (DRAS) that is used to facilitate the automation of customer response to various Demand Response programs and dynamic pricing through a communicating client. This specification, referred to as OpenADR, also addresses how third parties such as utilities, ISOs, energy and facility managers, aggregators, and hardware and software manufacturers will interface to and utilize the functions of the DRAS in order to automate various aspects of demand response (DR) programs and dynamic pricing." The OpenADR structure is illustrated in Figure 6, with the key features defined in Box 1.

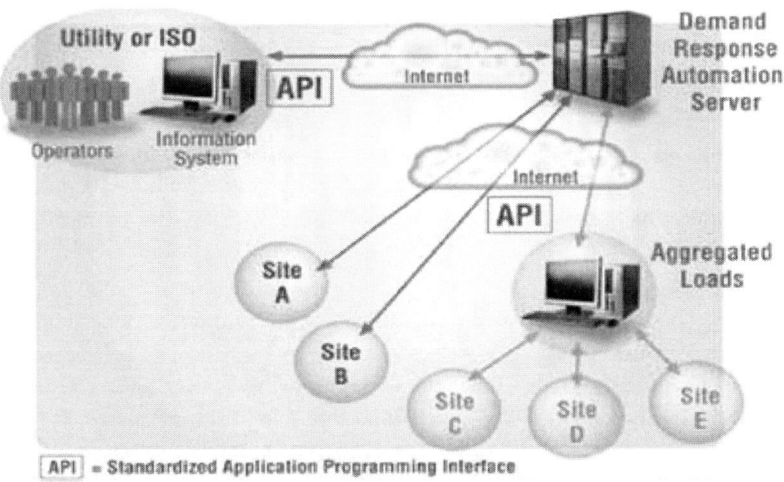

Figure 6. OpenADR Structure.

Box 1: OpenADR Features

Continuous, Secure and Reliable – Provides continuous, secure, and reliable two-way communications infrastructures where the clients at the end-use site receive and acknowledge to the DR automation sever upon receiving the DR event signals.

Translation – Translates DR event information to continuous Internet signals to facilitate DR automation. These signals are designed to interoperate with Energy Management and Control Systems, lighting, or other end-use controls.

Automation – Receipt of the external signal is designed to initiate automation through the use of preprogrammed demand response strategies determined and controlled by the end-use participant.

Opt-Out – Provides opt-out or override function to participants for a DR event if the event comes at a time when reduction in end-use services is not desirable.

Complete Data Model – Describes a rich data model and architecture to communicate price, reliability, and other DR activation signals.

Scalable Architecture – Provides scalable communications architecture to different forms of DR programs, end-use buildings, and dynamic pricing.

Open Standards – Open standards-based technology such as Simple Object Access Protocol (SOAP) and Web services form the basis of the communications model.

During a Demand Response event, the utility or RTO/ISO provides information to the DRAS about what has changed and on what schedule, such as start and stop times. A typical change would specify one or more of the following:

- Price signals: This would include a price multipler, a price relative, or an absoulte price
- Reliability signals: This would include the load amount to be shed (difference, load level, or set-point that a load should go to).
- Levels: These are simple representations of the price and reliability signals such as NORMAL, MODERATE, and HIGH.

The standard also specifies considerable additional information that can be exchanged related to DR and Distributed Energy Resources (DER) events, including event name and identification, event status, operating mode, various enumerations (a fixed set of values characterizing the event), reliability and emergency signals, renewable generation status, market participation.

Widespread adoption of OpenADR will accelerate the successful implementation of DR programs andDER, thereby providing the following four major benefits for all stakeholders:

- *Lower Costs* – Standardization lowers development and support costs for vendors and, ultimately, their utility customers. Standardization also fosters technology innovation and competition, which expands product choices for both utilities and end users.
- *Assured Interoperability* – Electricity providers and consumers alike benefit from being able to choose from among a wide range of different products and services without concern for any incompatibility or inevitable obsolescence.
- *Greater Reliability* – Products based on robust standards function dependably under normal circumstances and are able to recover from any anticipated error conditions to deliver dependable operation.
- *Enhanced Flexibility* – OpenADR has been designed to work with existing DR equipment (so-called backwards compatibility), as well as with newer, more sophisticated systems offering advanced feature sets.

Commercial, industrial and residential customers, and energy aggregators, will all be able to reduce costs, time and risk in the selection and deployment of products and systems based on the OpenADR standard. Work being performed by the OpenADR Alliance[26]will educate these stakeholders about the benefits of DR, and will increase their confidence in the available solutions with rigorous testing and certification programs.

As a result, equipment vendors and systems integrators will be able to accelerate the time-to-market for, and lower the development costs of, innovative products and services, while electric utilities, ISOs and RTOs will gain faster access to the market, experience lower capital and operational expenditures, and achieve greater success with DR programs. Even regulatory agencies will benefit from knowing that the introduction of new pricing policies will not be undermined by incompatibilities or other end-to-end impediments in the marketplace.

4.3. Smart Meter and Advanced Metering Infrastructure (AMI) and OpenADR

As the use of OpenADR for commercial and industrial facilities has gained significant traction in California and other parts of the U.S., the Advanced Metering Infrastructure (AMI) and smart home technologies are currently being implemented on a large-scale basis in residences. The AMI system wide implementation in California residences by the utilities together with development of the supporting technologies has provided opportunity for wide range of system operation and customer management applications, including communicating DR information through the AMI communication channels. The AMI communication is not open and accessible outside the utility network. AMI infrastructure can include smart meter, which is a revenue-qualified device from which charges can be derived. Other means of measuring power may be used, but they would generally not be qualified for revenue use from the residence point-of-view, the advanced meter contains valuable information about current and past power usage.

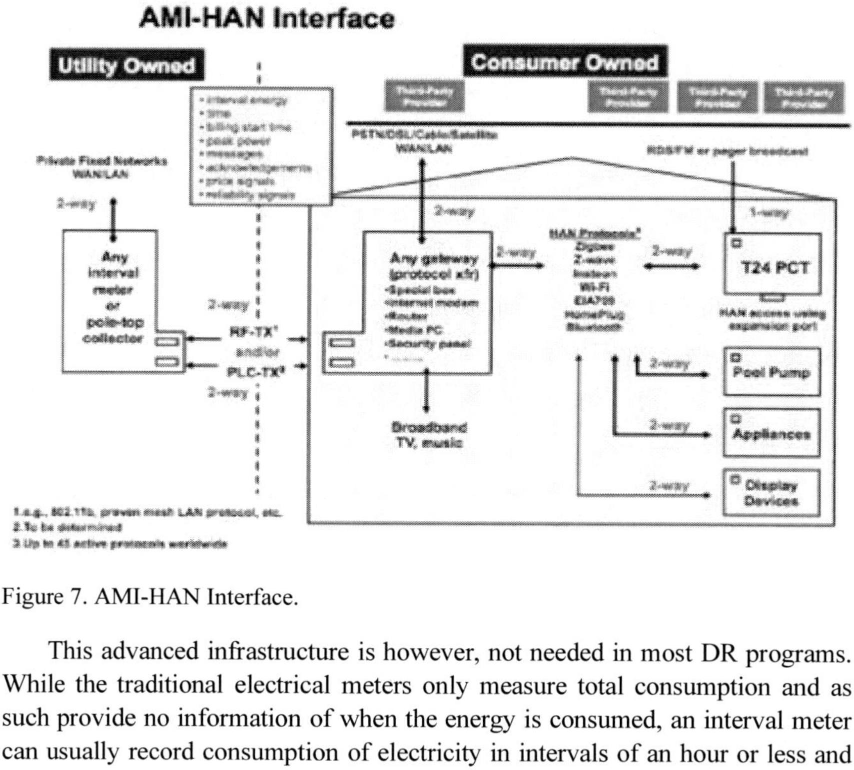

Figure 7. AMI-HAN Interface.

This advanced infrastructure is however, not needed in most DR programs. While the traditional electrical meters only measure total consumption and as such provide no information of when the energy is consumed, an interval meter can usually record consumption of electricity in intervals of an hour or less and communicate that information at least daily back to the utility for monitoring and billing purposes. In some DR programs, an interval meter is all that is needed for a customer to be qualified to participate.

LBNL is working with the utilities and other stakeholders to provide an external non-AMI based OpenADR interface to the residential technologies and home automation networks (HAN). These interfaces coexist with the AMI infrastructure that the utilities plan to use for their metering and billing purposes. Figure 7 shows these interfaces where OpenADR can be used as a means of communication directly with the residential gateway or the end-use devices such as the appliances:[27]

Further details on the home automation technologies and its use within the DR context are available from previous LBNL studies.[28]

5. Best Practices and Results of DR Implementation

The DR implementation best practices vary from different types of buildings. For example, a certain set of Heating Ventilation and Air-Conditioning (HVAC) strategies for commercial buildings may not be so relevant for a water and wastewater or data center industry. These best practices need to be validated through the actual results. Some of the best practices developed by LBNL are the result of over 10 years of research through pilots and development of DR strategies that are commercially implemented.29 [30] Since 2002, LBNL/Demand Response Research Center's (DRRC) research has been conducting research and development to advance DR technologies, policies, programs, strategies and practices. While most of these studies are focused on the commercial and industrial (C&I) sector, the DRRC has recently started research in residential sector. [31]

Over the years, DRRC has achieved significant success in accelerating the adoption of DR in various types of facilities and support DR program implementation successfully in California. As shown in Figure 8, this research includes development through field trials, standardization, and bringing technology to the market place.

The impact of the LBNL research has extended well beyond the state of California. Auto-DR is used in many other states in the US and is being piloted for deployments in other countries.[32] In this section, field experience, best practices and the achieved results from the research are provided.

Although, the primary goal of DR is to reduce electricity consumption during periods when wholesale price of electricity is high or when system reliability is jeopardized, the DR research has proven that DR has the capability of balancing the supply and demand at any time of the day.[33]

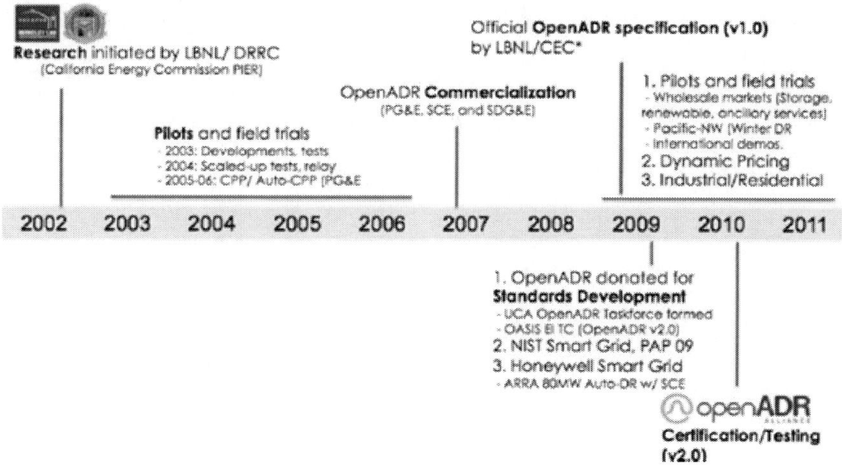

Figure 8. Roadmap of DR and AutoDR Research, Implementation and Commercialization.

As indicated in Section 4, DR can be implemented in a manual, semi-automated or fully automated fashion. Manual DR is labor intensive. Manual response can easily delay response or fail all together. Semi-automated DR means that the DR strategy is pre-programmed but requires a person to trigger it. 15% of the time people are not at the facility to respond to DR events.[34] Fully automated DR, or Auto-DR does not involve human intervention, and it is initiated by an EMCS in a facility through receipt of an external communications signal that triggers pre-programmed DR strategies. Automation helps improve the performance of DR by allowing the response to be more repeatable and reliable. Hence, whenever appropriate, Auto-DR may be a good practice moving forward.

5.1. DR Strategies in Commercial and Industrial Buildings

To provide reliable, repeatable DR, it is best to pre-program the strategy and fully automate the communication and controls to enable DR events without a human in the loop. This section discusses representative load reduction strategies and practices that can be used during DR events. These DR strategies work in well-tuned buildings when they are customized to sites' needs. In this section, we present DR strategies for commercial buildings and industrial facilities. Figure 9 identifies the typical end-uses that are in commercial and industrial buildings. It is from these end-uses that DR strategies are identified and acted upon.

Common End Use DR Loads		
Air handlers	Emergency generation	Motors
Anti-sweat heaters	Escalators	Outside signage
Chiller control	External lighting	Production Equipment
Chilled water systems	External water features	Processing lines
Cogeneration / CHP	HVAC systems	Pool pumps / heaters
Defrost Elements	Internal lighting	Refrigeration systems
Elevators	Irrigation pumps	Water heating

Figure 9. Common Examples of Commercial and Industrial End-Use Loads.

5.1.1. DR Strategies for Commercial Buildings

Heating, ventilation, and air conditioning (HVAC) systems can be excellent resources for demand reduction. First, HVAC systems comprise a substantial portion of the electric load in commercial buildings. Second, the heat storage of building envelopes and internal thermal mass allows HVAC systems to be temporarily unloaded without immediate impact on the building occupants. Third, a person feels comfortable with the indoor thermal environment within an allowable range of temperature, humidity and air speed. As long as these limits are not exceeded or exceeded for short periods, the DR strategies may be acceptable. Fourth, it is common for HVAC systems in large commercial buildings to be at least partially automated with energy management and control systems (EMCS).

Along with strategies for HVAC and lighting systems, turning off unnecessary plug loads is also commonly deployed in commercial buildings. In the case of automated DR (Auto-DR), all these strategies are usually pre-programmed into the facility EMCS that automate HVAC load shed for DR events. Otherwise, these strategies could be implemented manually or in a semi-automated manner. Curtailable loads also include discretionary loads such as outside signage, elevators, escalators, fountains and water features.

5.1.2. DR Strategies for Industrial Facilities

Implementing DR in industrial sectors presents a number of challenges, both practical and perceived. Some of challenges are: the wide variation in loads and processes across sectors and even within sectors; resource-dependent loading patterns that are driven by outside factors such as customer orders or time-critical processing (e.g. tomato canning); the perceived lack of control; and aversion to risk, especially unscheduled downtime. However, experiences in industrial sector have proven that with careful planning and preparation, the industrial sectors hold significant promise and great opportunities for DR. Here, a few specific examples of DR strategies are briefly described.

5.1.3. DR Strategies in Water or Wastewater Facilities

Wastewater treatment facilities are energy-intensive and have significant electricity demand during peak periods. Most of the facilities have storage ponds, which make load-shifting strategy possible. Turning off the energy-intensive equipment such as effluent pumps and centrifuges can result in significant load reduction.[35]

5.1.4. DR Strategies in Refrigerated Warehouse Facilities

Pre-cooling strategies, using the thermal mass in some of the refrigerated warehouses, enable load shifting from peak to off-peak periods. Shifting of batch process, shifting operation of equipment such as conveyors, pump systems, space conditioning, motors, process cooling, and storage can also be utilized. Another load shifting strategy is to defer forklift battery charging to off-peak hours.[36] Those load shifting strategies can be used in combination with load shedding strategies such as process shutdown; shutting down operation of equipment such as aerators, electrical, process air, shutting off air-handlers serving freezers, increasing set point on HVAC systems.

Refrigerated warehouses and cold storage facilities have also proven to be successful loads for the emerging field of bi-directional demand response. The ability to both increase and decrease end-user demand is becoming increasingly important with the rising penetration of intermittent renewable resources such as wind. For example, the Bonneville Power Administration (a large generation and transmission agency located in the Pacific Northwest region of the US) is currently operating a pilot program using loads for just this reason. The pilot program is using refrigerated warehouse and water pumping loads that can be altered in both directions for short periods of time, depending on the conditions of the wind system. If the wind is blowing during times of low system load, the loads in these facilities are increased. When the wind is not blowing, then the loads are decreased.

5.1.5. DR Strategies in Food Processing Facilities

Significant DR opportunities exist to both reduce and shift essential demand (i.e. manufacturing-related demand) and non-essential demand (e.g. office buildings, warehousing, etc.) in food processing facilities.[37] These strategies include adjusting operation schedules, adjusting raw material delivery, shutting off or adjusting setpoints of end-use applications, as well as adjusting lighting and HVAC systems. Various supply chain factors such as scheduling of raw material delivery, perishability, labor, logistics, shelf life, and product transport require the food processing sector in particular to carefully plan for curtailment or postponement.

5.1.6. DR Strategies in Data Centers

Data center are another energy-intensive sector. Because data centers are highly automated, they are excellent candidates for Auto-DR. However, their sophisticated controls of environmental conditions, high level of technology implementation, and users' technical knowledge make data centers' participation in DR unique, especially in implementing Auto-DR strategies. "Non-mission-critical" data centers are the most likely candidates for early adoption of DR. The largest opportunity for DR or load reduction in data centers is in the use of software algorithms, server consolidation and virtualization to reduce Information Technology (IT) equipment energy use, which correspondingly reduces facility-cooling loads.[38] In the case of data center located in multiple regions, load migration may work well for DR events. DR strategies could also be deployed for data center lighting, and HVAC systems. This is an ongoing area of research for the DRRC.

End-use curtailment is just one manner of response for data centers. More common is the utilization of on-site generation. Data centers and other mission critical facilities (e.g. financial institutions) are fully-backed with standby generation to ensure business continuity in the event of a grid emergency. These facilities often regularly test these generation units to ensure readiness should a power loss occur. Instead of testing at random times, these facilities can shift site loads onto these generation assets during periods of grid need, and achieve the double-benefit of testing under load and providing a resource.

5.1.7. DR Strategies in Heavy Industry

Heavy industrial and manufacturing is another well-represented vertical within C&I demand response. While such facilities often have very specialized equipment, their energy-intensive operations can represent significant load reduction capability. For slow response (i.e. 4 or more hour notification), production lines can often be fully shut down, and perhaps rescheduled. On a more rapid basis, variable speed drives, balers, and even arc furnaces can be both curtailed and remotely controlled. Of course, many industrial facilities also have standby and cogeneration capacity which can be leveraged as well.

6. Recommendations and Key Principles for Designing and Implementing DR in China

Based on the experiences in the US and elsewhere in the world, we offer the following recommendations for designing and implementing DR in China.

Provide Capacity Payments

Regardless of market structure, the most successful demand response programs are those that utilize capacity or availability payments and compensate participants for being ready and able to reduce load when called upon to do so. Such payments create a visible revenue stream allowing customers to better assess the costs and benefits of participation, and for DR providers and aggregators to invest in the requisite technology to ensure reliable performance.

Enable Meter Access

Reliable demand response depends on access to granular meter data. Whether interval meters with pulse outputs, or advanced meters with RF communication abilities, it is essential to ensure timely meter data access to participating customers and demand response providers.

Facilitate Accurate and Transparent M&V

One of the most important elements of DR is the specification of measurement and verification (M&V) procedures. One of the most important aspects of DR M&V is the calculation of the baseline, reflecting the load that would have been in place had there not been a DR program on the particular DR event day. Baseline methods are very important in that they form the entire foundation for assessing both the capacity delivered and financial payments warranted. A poorly designed baseline can have unintended negative consequences that can render participation unattractive and negatively impact uptake in the program or market. Recommendations for a baseline measurement include:

- DR resources should receive credit for no more and no less than the curtailment they actually provide. A baseline methodology should use granular meter data to create an accurate forecast of what load would have been used in the absence of a demand response dispatch.
- Baseline methods should not include attributes that encourage or allow customers to distort their baseline through irregular consumption nor allow them to "game" the system.
- The baseline should be simple enough for all stakeholders to understand, calculate, and implement. In addition, it should be possible to determine the baseline measure in advance of, or, during demand response disptaches so that DR providers can use the baseline measure to assess performance in real-time in order to ensure that delivery targets are met.

Encourage Aggregation

In other markets, specialized demand response providers, known as load aggregators, have contributed significantly to the growth of the industry and level of available DR resources. More than 125 utilities across the US are contracting with such outsourced DR specialists to implement C&I DR programs.39 Achieving similar rates of growth and levels of success in China without load aggregators would likely be quite difficult. Load aggregators provide a variety of benefits that contribute to DR uptake:

- Ability to match utility needs with customer capabilities.
  Across the various customer segments (e.g., residential, commercial, industrial), customers will have different capabilities to participate in a DR program. For example, some may be able to curtail their loads for as many as six hours at a time; others may be limited to shorter durations. In many cases, the capabilities of the specific C&I customers will not exactly match the needs of the utility. Load

aggregators can pair together customer sites in various ways to ensure that a specified level of capacity is always available to a utility.

- Core competency: Behind the meter expertise.
  There is significant potential to leverage the load reduction capabilities of large customers (e.g., commercial and industrial) as a capacity resource; however, in practice, realizing this potential requires a deep understanding of customer operations, since the "resource" is made up of dozens of decentralized assets. Unlike relatively homogenous residential loads, C&I loads vary widely across types of facilities.
  Effective load curtailment strategies must be customized to each organization's specific operational requirements to avoid impacting operations or occupant comfort. While smart-grid developments may facilitate two-way communications between the utility and its C&I customers, enabling those customers to effectively respond to price or control signals requires the implementation and ongoing management of individualized demand reduction plans. Most load aggregators employ professional engineers who have a deep understanding of electricity use and load patterns across a broad range of facility types, along with the experience to assess and identify strategies to successfully and consistently deliver DR capacity.
- Eliminates utility risk on performance.
  If demand response is considered to be part of a utility's overall resource portfolio, it is essential that a DR program delivers the amount of capacity expected each time the program is dispatched. The value of a DR resource to a utility is closely tied to its reliability. Utilities that operate internally-focused C&I DR programs such as interruptible tariffs typically assess penalties on customers that fail to perform during an event, as a means of mitigating customer non-compliance. This dynamic can put the utility in an uncomfortable position if a valued customer underperforms, given that the utility relationship with that customer extends well beyond the DR program. By contrast, DR service providers such as load aggregators are uniquely positioned to guarantee a reliable DR resource to a utility. Much like an insurance company reduces its exposure by pooling risks across a number of policy holders, DR service providers are able to assemble a portfolio of DR customers in a way that mitigates the risk of nonperformance for each of them. Notably, the larger the DR portfolio, the more reliable the resource will be.

- *Develop Open Standards.*
  OpenADR can be used to reduce the cost of DR programs and eliminate stranded assets. Creating an open, competitive market for controls vendors is an effective way to reduce the cost of DR equipment. Standards allow interoperability among systems. Open standards also facilitate the use of Auto-DR, which can be used to improve the performance of DR programs by allowing the response to be more repeatable and reliable. It is also essential when DR participates in ancillary services markets.
- *Create Opportunities for Training and Customer Education.*
  It is critical that training and education opportunities be created to help the customer understand the value of DR, the strategies and technologies needed to properly deploy DR, and the policies and procedures put into pact to enable further development of DR. Consumers need to be made aware of the time-varying value of electricity costs and take advantage of lower price periods.

End Notes

[1] U.S. Department of Energy (February 2006), Benefits of Demand Response in Electricity Markets and Recommendations for Achieving Them: A Report to the United States Congress Pursuant to Section 1252 of The Energy Policy Act of 2005, pp. 11-12. (http://eetd.lbl.gov/ea/ems/reports/congress-1252d.pdf).

[2] http://www.energy.ca.gov/2011energypolicy/documents/2011-07-06workshop/background /MetricsJulyIEPRDRv1.pdf.

[3] Synapse Energy Economics, "Modeling Demand Response and Air Emissions in New England," September 4, 2003. Page 16.

[4] EnerNOC, Inc. Analysis of US and Australian Electricity System Data.; Brattle Group, "The Power of Five Percent," May 16, 2007

[5] Brattle Group, "The Power of Five Percent," May 16, 2007.

[6] Id.

[7] U.S. Energy Information Administration (November 23, 2010), Demand-Side Management Actual Peak Load Reduction by Program Category. (http://205.254.135.24/cneaf/electricity /epa/epat9p1.html)

[8] US Energy Information Agency. 2009.

[9] Goldman, Charles, M. Reid, R. Levy, and A. Silverstein (January 2010), *Coordination of Energy Efficiency and Demand Response*, Lawrence Berkeley National Laboratory, LBNL-3044E. (http://eetd.lbl.gov/ea/ems/reports/lbnl-3044e.pdf)

[10] Peak load reductions are categorized as potential or actual. Potential peak load reductions are the amount of load available for curtailment through load control programs such as direct load control, interruptible load control, other load management, or other DSM programs. Actual peak load reductions are the amount of reduction that is achieved from load control

programs that are put into force at the same time as peak load and the amount of reductions that result from energy efficiency programs at the time of peak load.

[11] Id.

[12] Cappers, Peter, C. Goldman, and D. Kathan (2009), *Demand Response in U.S. Electricity Markets: Empirical Evidence*, Lawrence Berkeley National Laboratory, LBNL-2124E. (http://eetd.lbl.gov/ea/ems/reports/lbnl-2124e.pdf).

[13] The current size of the ISO-New England system is approximately 26 GW.

[14] The current size of the PJM system is approximately 165 GW.

[15] Federal Energy Regulatory Commission (2011). "Assessment of Demand Response & Advanced Metering Staff Report." November 2011. Table 2.

[16] Load aggregation is the process by which individual energy users band together in an alliance to secure more competitive prices than they might otherwise receive working independently. Oftentimes, load aggregator companies are formed to represent the interests of these groups of customers.

[17] Cappers, Peter, C. Goldman, and D. Kathan (2009).

[18] For example, the Fast Reserves (FR) program has a 50 MW minimum requirement for participation.

[19] The Reserve Capacity Cycle (RCC) is the process that is used in Australia to procure DR resources as part of the Reserve Capacity Mechanism.

[20] Note that to date this situation has never been experienced so no auctions have been called.

[21] U.S. Department of Energy (February 2006).

[22] U.S. Federal Energy Regulatory Commission (December 2008), *Assessment of Demand Response and Advanced Metering*, Washington D.C. (http://www.ferc.gov/legal/staff-reports/12-08-demand-response.pdf).

[23] http://collaborate.nist.gov/twiki-sggrid/bin/view/SmartGrid/OpenADR.

[24] Piette, M.A., G. Ghatikar, S. Kiliccote, E. Koch, D. Hennage, P. Palensky, and C. McParland. 2009. Open Automated Demand Response Communications Specification (Version 1.0). California Energy Commission, PIER Program. CEC-500-2009-063 and LBNL-1779E.

[25] The OpenADR Primer, White paper by the OpenADR Alliance (http://www.openadr.org/).

[26] http://www.openadr.org/.

[27] Figure courtesy: Ron Hoffman, California Energy Commission.

[28] McParland, Charles. Home Network Technologies and Automating Demand Response. LBNL, 2008. LBNL-3093E.

[29] S. Kiliccote, M.A. Piette, J. Mathieu, K. Parrish, Findings from Seven Years of Field Performance Data for Automated Demand Response in Commercial Buildings, Proceedings of the 2010 ACEEE Summer Study on Energy Efficiency in Buildings, Pacific Grove, CA, August 15-20, 2010, LBNL-3643E.

[30] Piette, M. A., G. Ghatikar, S. Kiliccote, D. Watson, E. Koch, and D. Hennage, "Design and Operation of an Open, Interoperable Automated Demand Response Infrastructure for Commercial Buildings", Journal of Computing Science and Information Engineering, vol. 9, issue 2, no. 2, 2009.

[31] http://drrc.lbl.gov/news/residential-energy-display-survey-reds-pilot.

[32] Kiliccote, S., M. A. Piette, and J. H. Dudley, Northwest Open Automated Demand Response Technology Demonstration Project, 2010.

[33] Dave Watson, Mary Ann Piette, Nance Matson, Sila Kiliccote, Janie Page (Lawrence Berkeley National Laboratory); Karin Corfee, Betty Seto, Ralph Masiello, John Masiello, Lorin Molander, Samuel Golding, Kevin Sullivan, Walt Johnson, David Hawkins (KEMA).

Automated Demand Response as a Grid Balancing Resource for the Integration of Renewables. Report for the California Energy Commission. In Press. 2011.

[34] Quantum Consulting Inc. and Summit Blue Consulting, LLC. Working Group 2 Demand Response Program Evaluation - Program Year 2004 Final Report. Prepared for Working Group 2 Measurement and Evaluation Committee. Berkeley CA and Boulder CO, December 2004.

[35] Thompson, Lisa; Lekov, Alex; McKane, Aimee; Piette, Mary Ann, Opportunities for Open Automated Demand Response in Wastewater Treatment Facilities in California – Phase II Report: San Luis Rey Wastewater Treatment Plant Case Study, LBNL3889E, August 2010.

[36] A. Lekov, L. Thompson, A. McKane, A. Rockoff, M.A. Piette, Opportunities for Energy fficiency and Automated Demand Response in Industrial Refrigerated Warehouses in California, LBNL-1991E, May 2009.

[37] Glen Lewis Group, Strategies to Increase California Food Processing Industry Demand Response Participation: A Scoping Study, LBNL-63668, November 2007.

[38] Ghatikar, Girish, M.A. Piette, S. Fujita, A. T. McKane, J.Q Han, A. Radspieler, K.C. Mares, D. Shroyer. Demand Response and Open Automated Demand Response Opportunities for Data Centers. California Energy Commission, PIER Program and Pacific Gas and Electric Company (PG&E), January 2010.

[39] Public Utilities Fortnightly, SPARK, "Outsourcing Negawatts," January 2010.

In: Reduced Energy …
Editor: Peter N. Ryan

ISBN: 978-1-62948-074-9

Chapter 2

DEMAND RESPONSE IN U.S. ELECTRICITY MARKETS: EMPIRICAL EVIDENCE*

Peter Cappers, Charles Goldman and David Kathan

ABSTRACT

Empirical evidence concerning demand response (DR) resources is needed in order to establish baseline conditions, develop standardized methods to assess DR availability and performance, and to build confidence among policymakers, utilities, system operators, and stakeholders that DR resources do offer a viable, cost-effective alternative to supply-side investments. This paper summarizes the existing contribution of DR resources in U.S. electric power markets. In 2008, customers enrolled in existing wholesale and retail DR programs were capable of providing ~38,000 MW of potential peak load reductions in the United States. Participants in organized wholesale market DR programs, though, have historically overestimated their likely performance during declared curtailments events, but appear to be getting better as they and their agents gain experience. In places with less developed organized wholesale market DR programs, utilities are learning how to create more flexible DR resources by adapting legacy load management programs to fit into existing wholesale market constructs. Overall, the development of open and organized wholesale markets coupled with direct policy support by the Federal Energy

---

* This is an edited, reformatted and augmented version of a Berkeley National Laboratory publication, LBNL-2124E, dated June 2009.

Regulatory Commission has facilitated new entry by curtailment service providers, which has likely expanded the demand response industry and led to product and service innovation.

1. INTRODUCTION

Demand response (DR) can be defined as: "Changes in electric usage by end use customers from their normal consumption patterns in response to changes in the price of electricity over time, or to incentive payments designed to induce lower electricity use at times of high wholesale market prices or when system reliability is jeopardized." [1, 2][1] This concept of demand response can be traced to the beginnings of the U.S. electric power industry (circa early- to mid-1890s), where system engineers and utility executives debated the optimal pricing regime for this new found service: Hopkinson's demand charge or time-of-day differentiated rates [6]. The universe of time-based retail rates has expanded significantly from these early days of the industry to now include real-time pricing (RTP), critical peak pricing (CPP) and variations thereof [2, 3, 7, 8].

As our definition suggests, U.S. utilities have also utilized incentive-based programs, often based on reliability differentiation, to elicit demand response from customers [3].

Utilities implemented load management (e.g. direct load control) programs and interruptible/curtailable tariffs in the early 1970s, both of which were in essence call options in which the customer sold the right but not the obligation for the utility to curtail or shed some of the customer's load in exchange for an upfront payment (in $/kW-month or a bill credit for participation) or a per kWh discount for the non-firm electricity consumption [2]. The initial interest in load management was driven in part by the increasing penetration of air conditioning which resulted in needle peaks and reduced load factors. With the advent of integrated resource planning in the late 1970s and 1980s, utilities increasingly recognized the system cost impacts of meeting peak loads and began to view load management as a reliability resource.

In the mid-1990s, with the advent of electricity restructuring, policymakers and utilities interested in facilitating the development of regional, competitive wholesale (and, in some states, retail) electricity markets initially focused primarily on market design and structure, albeit with a supply-side focus (e.g., open access to transmission services, vertical de-integration, establishing independent system operators). However, the problems in many restructured electricity markets (e.g. electricity crisis in Western state power markets in 2000-2001, price volatility and spikes, perceived market power, reliability concerns during system peak demand conditions, and failure to produce expected benefits to consumers) led policymakers to conclude that demand response, in all of its different forms, is essential to the efficient functioning of wholesale electric markets [9]. The Energy Policy Act of 2005 (EPACT) codified that a key objective of U.S. national energy policy was to eliminate unnecessary barriers to wholesale market demand response participation in energy, capacity, and ancillary services markets by customers and load aggregators, at either the retail or wholesale level.

It is therefore critical to assess existing capabilities of DR resources among load serving entities and customers to provide load reductions in response to system emergencies and/or high market prices, and assess the actual performance of DR resources during recent periods. Empirical evidence of DR resources is needed in order to establish baseline conditions, develop standardized methods to assess DR availability and performance and to build confidence among policymakers, utilities, system operators, and stakeholders that DR resources do offer a viable, cost-effective alternative to supply-side resources. In this study, we summarize the existing contribution of DR resources in U.S. electric power markets (i.e., retail and wholesale), with a primary focus on enrollment and performance of incentive-based DR programs in organized markets (rather than time-based retail rates). Both types of DR resources are critical to the development of competitive electricity markets [10 - 12].

This paper proceeds as follows. First, it provides an overview of the types and magnitude of existing DR resources in the United States and then focuses on the evolution and maturation of incentive-based DR programs in organized markets in terms of enrollment and performance. Next, it discusses the evolution of legacy, existing load management programs and interruptible/curtailable tariffs offered by utilities within the new framework of organized wholesale markets, drawing on results of recent studies of the Midwestern Independent System Operator (MISO) and Southwest Power Pool (SPP) conducted by the authors. Finally, it explores the role that third party DR program providers (i.e. curtailment service providers) have played in expanding the scope of the DR industry, again drawing on the empirical evidence of recent activity.

2. CURRENT SIZE AND SCOPE OF DR IN THE UNITED STATES

As part of the Energy Policy Act of 2005, the U.S. Congress directed the Federal Energy Regulatory Commission (FERC) to develop a comprehensive national assessment of the size and scope of electricity DR resources and advanced metering as part of a national energy policy [2]. To accomplish this task, the FERC prepared and administered a comprehensive survey, first in 2006 [2] and then again in 2008 [5], to ~3300 organizations representing all aspects of the electric delivery industry (e.g., investor-owned utilities, municipal utilities, rural electric cooperatives, power marketers, state and federal agencies, and unregulated DR providers) from all 50 states. About 55% of these organizations (~1850 responses) completed the DR section of the survey.

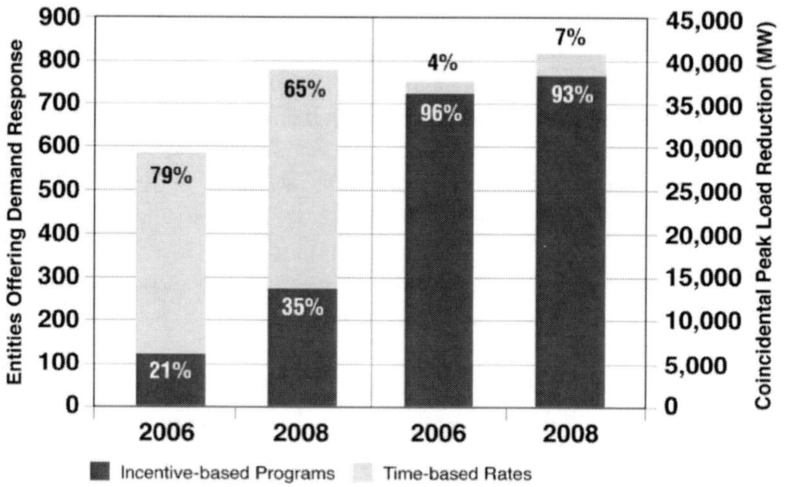

Source: [2] and [5].

Figure 1. Estimated size of DR resources in the United States.

Among survey respondents, there has been a significant increase (117%) in the number of entities offering DR programs: 126 in 2006 vs. 274 in 2008 (Fig. 1) and about a 10% increase in the number of entities offering dynamic pricing tariffs to retail customers. Nationally, the potential size of peak load reductions from existing DR resources, relative to national peak demand, was about 5.0% in 2006 [2] and grew to 5.8% in 2008 [5].[2]

Many more entities offer some type of time-based retail rate as compared to incentive-based DR programs. However survey respondents indicated that these time-based retail rates account for a small part of the total existing DR resource base. In 2008, customers enrolled in existing incentive-based DR programs were capable of providing ~38,000 MW of potential peak load reductions, while time-based retail rates were expected to produce another 2,700 MW (Fig. 1). In percentage terms, about 93% of the peak load reduction from existing DR resources in the U.S. is provided by various types of incentive-based programs (Fig. 1).

Given that peak loads vary significantly by region, it is also useful to characterize existing DR resources compared to a region's summer peak demand (see Fig. 2). Demand response resource potential ranges from 3 to 9% of a region's summer peak demand in most regions, with the notable exception of the Midwest Reliability Organization (MRO) region where DR resources represent a much higher percentage of summer peak demand.

Several factors may help to explain this result: (1) several states (Minnesota and Iowa) require utilities to invest a percentage of revenues from retail sales (1.5-2%) in demand-side management (DSM) programs, (2) utilities in the upper Midwest have historically had favorable resource adequacy rules that allow load management to be counted towards meeting reserve requirements, and (3) the customer base includes a significant fraction of industrial load that is amenable to interruption (e.g. steel plants) [5]. Among the existing DR resource base, residential customers account for ~6,000 MW while industrial customers account for ~14,800 MW. In the Midwest Reliability Organization (MRO), Northeast Power Coordinating Council (NPCC), and Reliability First (RF) regions, a significant portion of DR resources is attributable to programs offered by ISOs and RTOs (e.g., classified as wholesale). Elsewhere in the U.S., the majority of existing DR resource potential comes from more traditional DR programs: interruptible/curtailable rates for industrial customers and direct load control for residential and small commercial customers.

3. DR Resources in Eastern U.S. ISOs

3.1. Participation

Most of the growth in incentive-based DR resources has occurred in organized wholesale markets administered by ISOs/RTOs. Since 2001, FERC has required ISO/RTOs to file annual program evaluations or include a detailed discussion of their DR program enrollment and performance in annual state of the market reports. To illustrate trends in the development of DR resources in some organized markets, we focus on three ISOs located in eastern U.S. electricity markets: New York Independent System Operator (NYISO), ISO New England (ISO-NE), and PJM Interconnection (PJM).

The New York ISO has historically offered three different incentive-based DR programs: Emergency Demand Response Program (EDRP), Special Case Resource (SCR) program, and the Day-Ahead Demand Response Program (DADRP). EDRP is a voluntary program that pays strictly for energy; while SCR provides an up-front payment for capacity, a payment for load reductions when dispatched, but includes the threat of penalties for noncompliance with capacity obligations during declared program events.[3] The DADRP is an economic DR program that allows participants to submit load curtailment (i.e., supply) offers into the NYISO's Day-Ahead Market, where they compete side-by-side with generators. If a DADRP participant's offer is accepted, that participant is obligated to curtail the committed amount the following day, or else covers any open position it has at the higher of the real-time or day-ahead location-based marginal price (LBMP).

ISO-NE also offers three incentive-based DR programs: its Real-Time Demand (RTDemand), Real-Time Price (RT-Price) and Day-Ahead Load Response (DALRP) program. There are three options for customers wishing to participate in ISO-NE's emergency (RT-Demand) DR program: the first two require participants to send near real-time meter data every 5-minutes to the ISO, but differ in terms of the length of notification prior to an event they require (i.e., RT-30 Minute or RT-2 Hour) and consequently the floor energy price paid for curtailments ($500/MWh and $350/MWh, respectively); the third option (RT-Profiled) requires neither communications devices nor interval meters to be installed in order to participate. In all the cases, enrolling participants are subject to non-performance penalties. The RT-Price program provides customers the opportunity to reduce load in real-time when a specific price point is exceeded, while the DALRP offers customers the opportunity to participate indirectly in the ISO-NE's Day-Ahead energy market.[4]

PJM provides its customers with three incentive-based DR programs: Emergency, Active Load Management (ALM), and Economic load response programs. The Emergency and ALM programs are dispatched under system emergencies, but differ in terms of the requirements for participation. As a result of the introduction of a forward capacity market in 2007, the design of the Emergency and ALM programs was altered to accommodate these respective resources in the Reliability Pricing Model (RPM).[5] The Economic program has given customers the opportunity to participate in the Real-Time energy market, either through direct or indirect scheduling.

In addition, these three ISO/RTOs have recently developed opportunities for DR resources to participate in ancillary service markets. Both the NYISO and PJM allow DR resources to participate in regulation, 10-minute and 30-minute operating reserves markets. ISO-NE is offering a pilot program for customers to participate in providing operating reserves.

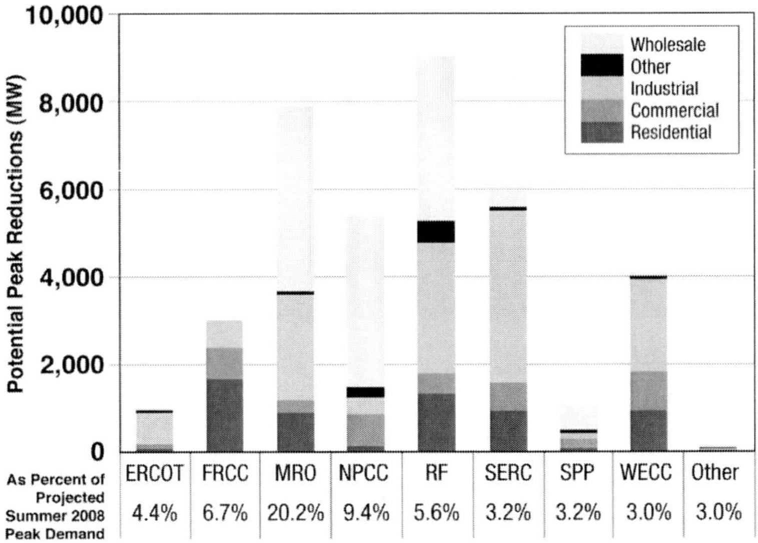

Source: [5].

Figure 2. Estimated size of DR resources by NERC region and customer sector.

Fig. 3 illustrates how DR program enrollment has evolved in these three ISO/RTOs from 2002 through 2007. In New York, the total size of the DR resource portfolio has not changed dramatically over the past six years, although there have been significant changes in the mix of individual programs.[6] Since 2003, when joint participation in EDRP and SCR was no longer allowed, there has been a clear migration away from EDRP towards the more lucrative but demanding SCR program.[7] The up-front reservation payment provided in the SCR program provides an ongoing revenue stream that is crucial to the financial viability of load aggregators and attractive to customers.

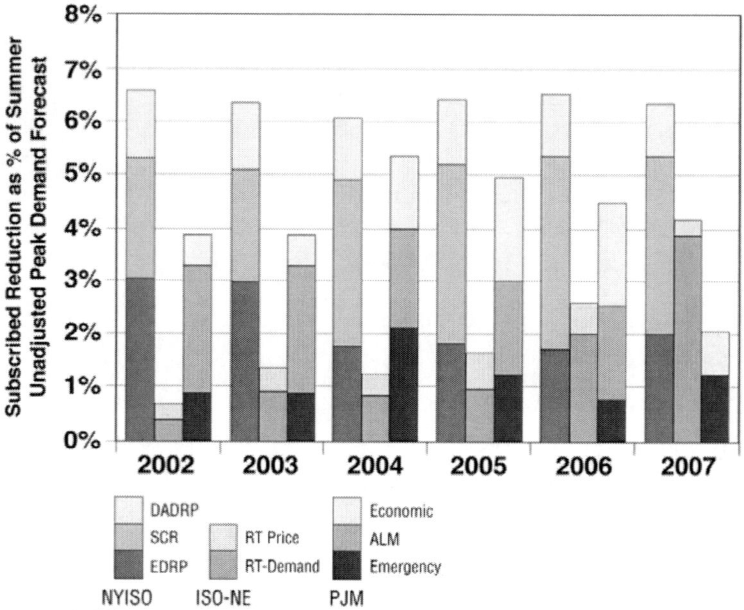

Source: FERC filings, annual State of the Market reports, working group presentations.

Figure 3. Comparison of Northeastern ISO incentive-based DR program enrollment.

In interpreting results for PJM, it is important to recognize that PJM significantly expanded its footprint (and summer peak demand) since 2005 and 2007 as new utilities from the Midwest joined PJM. Enrollment in PJM's DR programs has grown significantly from 2002 to 2007 from ~2100 MW to 4600 MW, although in percentage terms it is lower because PJM's footprint has expanded. PJM has also seen a major shift in its pool of program participants over time.

Significant changes in the designs of the Emergency and ALM programs were undertaken as part of Reliability Pricing Model (RPM), development process, in order to allow as many resources as possible to participate in that forward capacity market. Such alterations in the programs' designs appear to have elicited an exodus from the purely voluntary Emergency program option (i.e., Energy-Only) towards the capacity-based (Full and Capacity-Only) options, the latter labeled as ALM in Fig. 2.

During this period, participants enrolled in PJM's economic DR programs increased their subscribed peak load reductions from 335 MW in 2002 to ~2500 MW in 2007. Customers in two zones, Commonwealth Edison

(ComEd) and Baltimore Gas & Electric (BGE), account for nearly 65% of this increase in the economic program's capabilities. Since joining PJM, ComEd has enrolled and transitioned its existing DR assets into PJM's economic DR programs and also expanded its offering of DR programs (i.e., Early Advantage, Rider 32 and Voluntary Load Reduction Programs) [13]. Enrollment in economic DR programs in the BG&E zone almost tripled in 2007 compared to 2006 (140 to 393 MW), which may have been a response to very large rate increases in Maryland and aggressive marketing by curtailment service providers (CSPs).

In ISO-NE, the RT-Demand response program increased in size from 0.4% of forecasted system peak demand in 2002 to 3.9% in 2007, nearly an 800% increase in just 5 years. Since 2007, the introduction of the Forward Capacity Market (FCM) in New England has also contributed to the continued growth in DR resources, as estimated peak load reductions associated with the RT-Demand response program increased by 51% between 2007 and 2008.

Enrollment in DR programs provides system operators with an indication of the size of the customer resource base that is willing to curtail or shift load in response to system contingencies or high market prices. However, because participation is voluntary in some of these DR programs and because the utility often does not have physical control of the customer's load response (as in a direct load control program), information on the actual performance of DR resources during system emergencies or in response to high prices is crucial to assessing the long-term viability of DR resources. In order for these resources to be treated comparably to "iron-in-the-ground" generation assets, system operators must be confident that DR resources will perform in a consistent and predictable fashion. Performance metrics offer all market participants, and especially system operators, the opportunity to tangibly recognize the value of DR.

3.2. Performance

Two different performance metrics have been proposed by evaluators of the NYISO DR programs: Subscribed Performance Index (SPI) and Peak Performance Index (PPI) [14]. SPI compares the actual load reduction to what was initially subscribed to a DR program, while PPI estimates the customer's actual DR load curtailment compared to their peak demand. Given the infrequent reporting of the PPI by ISOs and the difficulty of producing the PPI independently, we focus on the SPI. For consistency of reporting, we focus on the portfolio level metric, whose definition was taken from [14]:

$$SPI_p = (E_d / E_s) \cdot 100\% \tag{1}$$

where,

$$E_d = \sum_{j=1}^{M} \left(\sum_{t=1}^{N} (CBL_{i,t} - E_{i,t}) \right) \tag{2}$$

$$E_d = \sum_{j=1}^{M} \left(\sum_{t=1}^{N} (CBL_{i,t} - E_{i,t}) \right) \tag{3}$$

and

Ed = the total electric energy curtailment *delivered* by all customers in a program,

E_s = the total electric energy curtailment *subscribed* by all customers in a program, *CBLt* = the customer baseline of customer *i* in hour *t* (MWh),

$E_{i,\,t}$ = the electric energy of customer *i* in hour *t* (MWh),

M = the total number of customers in a program,

N = the number of hours per curtailment event, and

$E_{sub,i}$ = the subscribed load curtailment of customer *i* (MWh).

We were able to derive SPI values in Fig. 4 from ISO-NE and NYISO DR programs for several years based on the evaluation results or reported program performance. In some cases, the lack of a reported metric in certain years is either because no events were declared (e.g., 2004) or conditions surrounding a declared event would not produce an accurate assessment of performance relative to subscription (e.g., 2003 Northeast Blackout).

The average SPI for ISO-NE's Economic program is 0.32 for four years with program data, which suggests that program participants' load curtailments were only about 32% of their subscribed load commitment during high price events.

The SPI varied considerably from one year to the next – ranging between a low of 9% (2002) and a high of 53% (2003), illustrating how highly variable performance is of these enrolling participants.

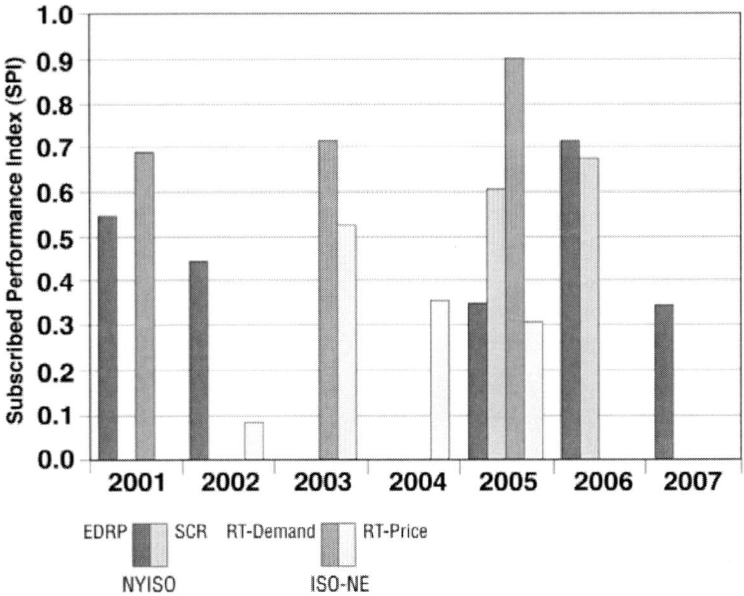

Source: FERC filings, annual State of the Market reports, working group presentations.

Figure 4. Comparison of Northeastern ISO program performance.

The relatively low and highly variable SPI for the ISO-NE Economic DR program is not too surprising, given the fact that this program was new to participants (who may have been unsure about how much load curtailment to "subscribe"), there were no penalties for non-performance, and participants had complete discretion concerning when and how much of a load reduction to undertake based on an economic analysis of the opportunity cost of consuming load. In our view, this type of performance metric can provide useful information over time (as customers obtain more experience with measurement and verification protocols used to estimate curtailed load during events) and if training and technical assistance are provided to customers to help them quantify the amount of discretionary load that they can and are willing to curtail or shed during events.

The two DR capacity market programs (NYISO SCR and ISO-NE RT-Demand) provided 64% (SPI=0.64) and 77% (SPI=0.77) of their expected curtailments, respectively. The voluntary emergency DR program (NYISO EDRP) produced an overall average SPI of 0.52. These results suggest that the actual performance of DR programs with noncompliance penalties will be closer to their committed load curtailment compared to economic or voluntary emergency DR programs (that do not have penalty provisions). In terms of consistency, the NYISO SCR and the ISO-NE RT-Demand programs' performance index also varies considerably less than the SPI for EDRP. The variability in SPI over time is also very important to system operators who have the responsibility for maintaining grid reliability.

If DR is to play an increasing role in wholesale markets as an economic or reliability resource, system operators and resource planners must be able to accurately predict what DR resources can provide during system events in order to maximize their contribution to market efficiency and system stability while minimizing overall system costs. Recent efforts by system operators to manage and integrate intermittent generation resources provide an instructive example. Intermittent generation resources, like DR, are playing an increasing role in the bulk power system. As wind generation has grown over the past several years, ISO/RTOs have been forced to not just rely on these resources' accepted offers in forward markets to predict real-time performance but have also developed internal forecasts of their output in order to ensure sufficient reserves are in place to maintain reliability. For example, in September 2008, the NYISO brought on-line a new state-of-the art wind forecasting system that feeds wind-power forecasts based on meteorological data and historical operating characteristics directly into NYISO operational systems to better maintain the requisite balance of load and generation and predict wind power output on an hourly basis [15].

4. Integration of Existing Utility DR Programs in Wholesale Markets

As part of the transition to competitive, organized wholesale markets, it is necessary for the wholesale market rules and requirements to accommodate and facilitate a transition of existing DR resources into these new markets. Initially, the design of organized wholesale markets focused primarily on developing market rules that worked for supply-side assets. The FERC and state regulators in a number of states have placed increasing emphasis on ensuring that market rules provide an opportunity for existing DR assets enrolled in legacy incentive-based programs to participate in organized wholesale markets.

Working with the Lawrence Berkeley National Laboratory, the Organization of Midwest's Demand Response Initiative (MWDRI)[8] and the

Southwest Power Pool each commissioned a detailed survey of the design features, operational triggers used to call events (e.g., system emergencies, market conditions, local emergencies), DR resource availability (e.g. seasonal, annual), participant incentive structures, and historic performance of existing DR programs and dynamic pricing tariffs offered by load serving entities in each ISO/RTO [16 - 17]. Although the timing of the surveys differed by roughly a year, they shared common goals:

- To inventory the existing set of retail incentive-based DR programs and dynamic pricing rates;
- To assess differences and similarities among existing retail incentive-based DR programs and dynamic pricing rates; and
- To help inform the debate at MISO and SPP concerning how to best make use of existing retail DR assets at the wholesale level.

The survey for SPP was fielded to 52 different cooperatives, municipal utilities, investor-owned utilities (IOU), state agencies, independent power producers (IPP), power marketers and transmission companies. Thirty entities, all municipal utilities and IOUs returned completed surveys; 14 of these entities offered some form of DR to their customers. In the MWDRI survey project, 35 utilities completed the survey with information on 141 DR programs and dynamic pricing tariffs; survey response was very good (~80%).[9] In terms of wholesale market design, SPP administers an Energy Imbalance Service (EIS) market: participation is mandatory for load serving entities and generators and all real-time resources where imbalances are settled using the EIS market.

Table 1. Overview of SPP and MWDRI survey results

	SPP			MWDRI		
	Incentive-based programs	Time-based rates	Voluntary response	Incentive-based programs	Time-based rates	Voluntary response
Survey Respondents	26	5	4	99	12	N/A
No. of Programs	36	5	6	122	19	N/A
Potential Coincident Peak Demand Reduction (MW)	1,352	200	N/A	4,406	321	N/A
Distribution of DR Resources	87%	13%	N/A	93%	7%	N/A

Source: [16] and [17].

MISO administers a day-ahead and real-time energy market with centralized economic dispatch and locational marginal pricing as well as ancillary services markets for regulation, spinning reserves and supplemental reserves. At the time of the survey, neither ISO/RTO had explicit wholesale DR programs that the ISO/RTO administered. As Table 1 illustrates, both SPP and MISO have a robust existing set of DR resources capable of reducing RTO system peak demand, roughly 4% and 5% respectively. Specifically, the SPP survey revealed that in 2008 there were a total of 47 different retail DR initiatives currently being offered in that RTOs footprint: 36 incentive-based DR programs, five time-based retail rates, and six voluntary response programs.[10] Retail DR as a whole in SPP is estimated to provide 1,552 MW of potential coincident peak demand reduction, 13% of which comes from customers on time-based rates while the remaining 87% is associated with incentive-based DR programs.[11] The MWDRI study indicated that as of late 2007 there were over 122 different retail incentive-based DR programs being offered to customers, and 19 different time-based retail rates in the region's utility tariffs. Collectively, DR is forecasted to reduce coincident peak demand in MISO by 4,367 MW, again the vast majority (93%) of which is coming from incentive-based DR programs. The surveys also provided insights into how utilities in SPP and MISO are utilizing their DR resources. Respondents were asked to characterize the conditions (i.e., improving local reliability, mitigating system emergency conditions, and/or reducing exposure to high market prices) under which they chose to invoke load curtailments.

Historically, interruptible/curtailable (I/C) and direct load control (DLC) programs were justified primarily for reliability purposes and dispatched only during system emergencies. However, as competitive wholesale markets have developed and with the formation of MISO and SPP, most DR programs in these two regions currently have more than one operational trigger. A Venn diagram illustrates the universe of different conditions under which program administrators are invoking their DR resources and the expected load reductions associated with each combination of conditions.[12] The dark intersections in Figs 5 and 6 shows that in SPP and OMS states, about 69% and 64% respectively of survey respondents' enrolled DR (in MW), can be called for multiple conditions. An increasing number of utilities are now recognizing the flexibility these tariffs provide in the new wholesale market environment by also allowing for economic dispatch of these DR programs.

When utilities see or expect high prices, at least 66% of the peak demand reductions associated with a DR program (i.e., DLC or I/C) in SPP and OMS can be (or have been) dispatched (see Figs. 5 and 6). Utilities in OMS indicated that they wanted to reduce their exposure to high energy market prices but were reticent about bidding these resources into MISO's day-ahead market directly. So instead, the utilities themselves dispatched these programs closer to real-time when energy market prices rose above a certain level. In contrast, the distribution cooperatives in SPP who responded to the survey invoked their DLC programs for flattening out their load shape in order to minimize coincident transmission system peaks, thereby achieving substantial savings in their demand charge.

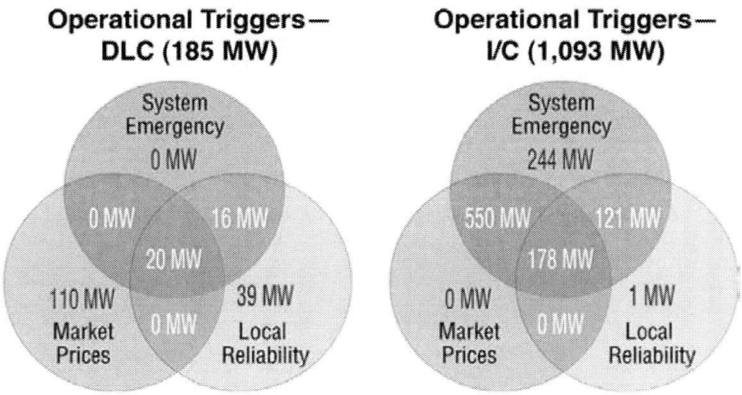

Source: [17]

Figure 5. Peak load reduction by operational trigger for DLC and interruptible DR programs in SPP.

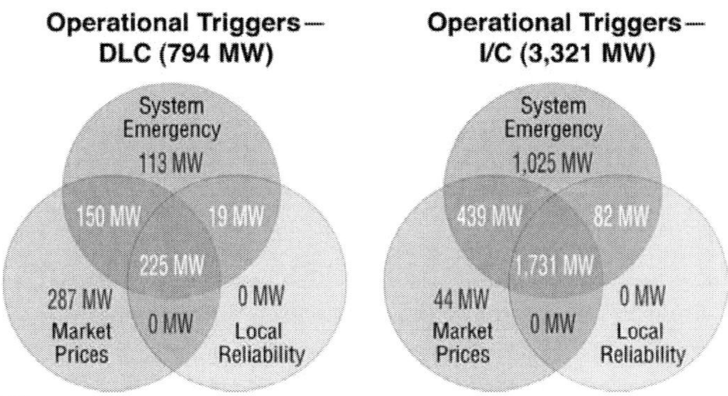

Source: [16].

Figure 6. Peak load reduction by operational trigger for DLC and interruptible DR programs in OMS states.

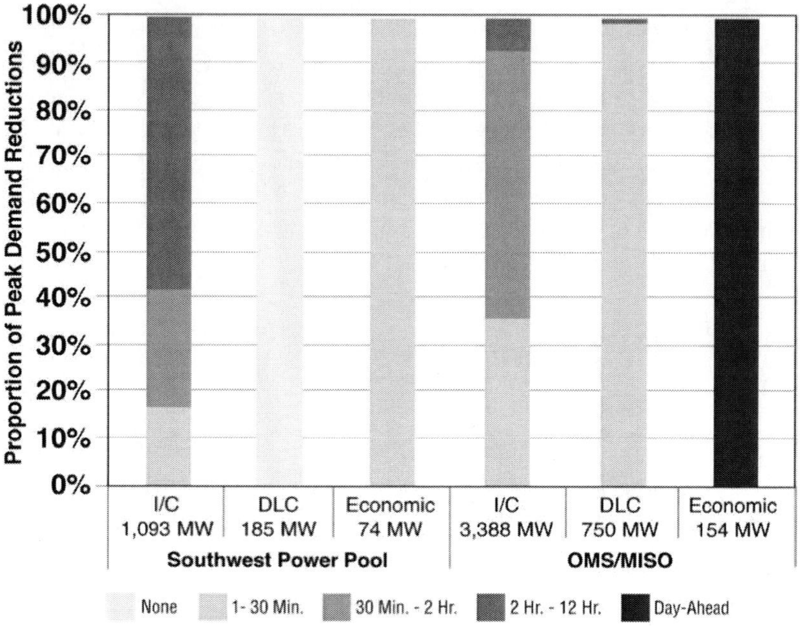

Source: [16] and [17].

Figure 7. Advance notification requirements for DR Programs.

The surveys also requested that respondents specify the number of hours of advance notice required before demand can be reduced for each DR program. Advance notice requirements vary considerably across DR programs and by region (see Fig. 7). For example, DLC programs were uniformly reported to have no or very short notice requirements, which is not surprising given that equipment is cycled directly by utilities. In contrast, for interruptible/curtailable tariffs in MISO, ~90% of the enrolled load could be curtailed in less than 2 hours of advance notice with a significant amount of that load (1960 MW) available on just 30 minutes notice. However, among SPP member utilities, survey respondents reported that only 40% of the enrolled interruptible/curtailable load could be curtailed within 2 hours (see Fig. 7). Economic DR programs do not have significant amounts of enrolled load in either SPP or MISO, although notice requirements are much shorter in SPP (1-30 minutes) compared to MISO (day-ahead).

The relatively short event notification requirements associated with DLC programs make them perfect candidates to participate in wholesale real-time ancillary services markets, when such opportunities arise. If emergency and/or capacity DR programs are developed at SPP and/or MISO, then the vast majority of I/C resources could participate under existing retail program and tariff structures.

5. Role of Curtailment Service Providers in Wholesale Market DR Programs

One of the arguments and intended benefits of competitive wholesale markets was service and product innovation. The emergence and increasing role of curtailment service providers provides an interesting case study that illustrates how strong public policy support by FERC and stakeholder support in organized wholesale markets created opportunities for new entrants to obtain a significant foothold and thus expand the DR industry.

In virtually all ISO/RTOs, "legacy" incentive-based DR programs offered by utilities were the initial participants in wholesale market DR programs. In states with retail competition, it was not long before competitive (non-utility) entities began offering customers similar opportunities. Each of the Eastern ISOs (ISO-NE, NYISO, and PJM) that developed opportunities for end-use customers to participate in their wholesale markets had to develop market and program rules for load aggregators that proposed to offer a customer's load reduction capability as a paid resource but were not the customer's load serving entity (LSE).

Program design and implementation issues that had to be addressed in order to facilitate participation by CSPs included: (1) a more sophisticated registration process for load aggregators (e.g., ensuring that customers' sites were not enrolled by multiple program providers), (2) notification procedures (e.g., notifying load serving entities that customers were enrolling in a incentive-based wholesale market DR program by a CSP), (3) metering and telemetry requirements (e.g., access by CSPs to customer's interval meter data) and (4) back-office software modifications at the ISO and incumbent utility in order to ensure timely and accurate processing and transmission of the interval data to CSP and ISO.

As new entrants, CSPs incurred substantial up-front costs which included marketing costs to enroll customers in a DR program, back-office and communications network infrastructure costs, and design, installation, financing, and maintenance of enabling technology at customer facilities (e.g. controls, onsite generation). CSPs required a source of revenue to make program participation a viable business opportunity. Energy payment for verified load reductions achieved by enrolled customers was an option, although CSPs would have to rely on the likelihood that events would be called by an ISO, which could be problematic. In contrast, utilities were typically allowed to recover program administration costs directly into retail rates.

CSPs soon gravitated towards incentive-based DR programs (e.g., capacity market, requests for emergency resources) that provided an upfront and ongoing reservation payment for committed load reduction by load aggregator (or customer). These programs provided a significant opportunity for CSPs to aggregate individual customer's willingness to curtail into a load curtailment resource, negotiate and share reservation payments with customers, provide energy payments to customers for performance during events, and allow CSP to compete on the basis of price and not just service. For example, Fig. 8 shows enrollment by the type of service provider (i.e., CSP or utility) in several DR programs administered by the NYISO. From 2003 to 2008, CSPs increased their share of subscribed load of DR resources from 44% to 77% in the emergency (EDRP) and capacity markets (Installed Capacity/Special Case Resources – ICAP/SCR). The ICAP/SCR program has been the main area of growth for CSP, accounting for well over 80% of the enrolled capacity in 2008. CSPs have heavily marketed the SCR program to customers by developing customized service packages and enabling technology that help customers to manage the risks associated with participation. Enrollment in the voluntary EDRP program has steadily eroded (i.e., 956 MW in 2003 but only 365 MW in 2008) as CSPs have shunned the EDRP program that provides energy payments only during events. The market share of utilities has steadily declined over this period.

CSPs also have been successful in attracting new customers to enroll and participate as DR resources in wholesale market DR programs. Results from ISO-NE's Forward Capacity Market auction illustrate this phenomenon. In 2007, ISO-NE filed with FERC its approved Forward Capacity Market (FCM) rules, which would allow any resource, both supply and demand, to commit three years ahead of time to provide capacity to the system [18]. Demand resources in the FCM included both DR and energy efficiency, and load aggregators had to identify if the resource already existed (e.g., generator currently producing electricity, end-use customer currently enrolled in a DR program) or was new (e.g., planned generation addition, expected future enrollment in a DR or energy efficiency program).

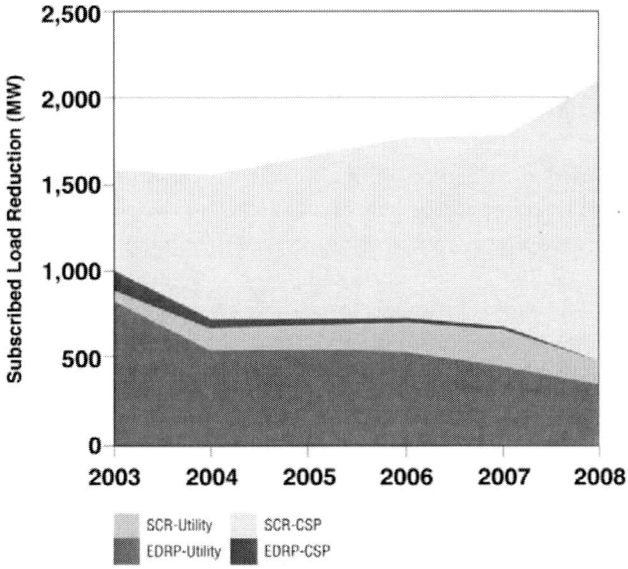

Source: FERC filings, annual State of the Market reports, working group presentations.

Figure 8. NYISO DR program enrollment: utilities vs. CSP.

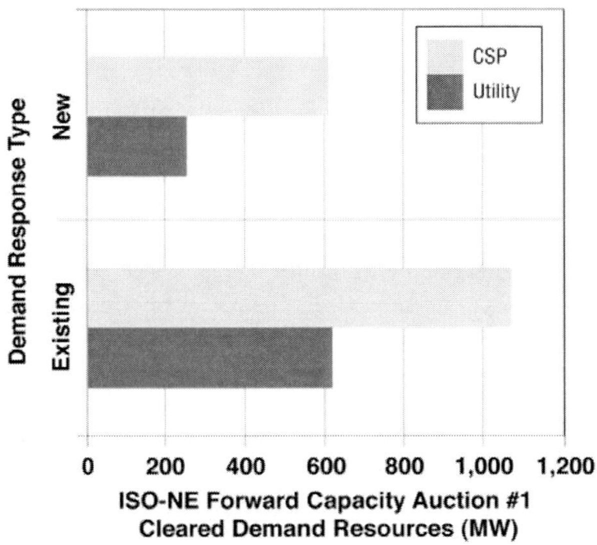

Source: [19].

Figure 9. Distribution of cleared demand-side capacity in ISO-NE FCA #1.

The results of the first Forward Capacity Auction (FCA #1) were made public in March 2008 [19].[13] Across the six New England states, CSPs were responsible for attracting over 60% (1,681 MW) of the total demand-side capacity (2,553 MW) that cleared in the FCA #1 and 70% of the new demand-side resources (see Fig. 9). These results suggest that CSP were more aggressive in marketing and/or willing to take the business risk that they could deliver demand resources three years hence.

CSP still face significant institutional and regulatory barriers in many regions of the United States. For example, some states (e.g., Indiana) have precluded third party program providers or customers from directly participating in wholesale market DR programs. Many Public Utility Commissions (PUCs) also limit the share of program benefits that may be retained by the utility, opting to give the bulk of them back to consumers (e.g., New York). With CSP, the sharing of benefits is typically part of the contract negotiation process. PUCs are also concerned about the erosion of their authority to regulate the business and operations of incumbent monopoly utilities and its infrastructure. Some states have argued that they have a legitimate reason for not opening up their retail sector to aggregators of retail customers (or "ARCs") and such decisions should be respected. The FERC has attempted to finesse this issue in its recent Order 719 [20] in which the FERC agreed with the principle that load aggregators must be allowed to participate in ISO/RTO markets unless prevented under state law or regulation. However, FERC did not make it clear who was responsible for notifying the ISO/RTO that a state precluded customers from participating in wholesale DR programs with a non-utility entity.[14]

Traditionally, DR vendors provided load control and communication/ notification technologies to utilities on a fee-for-service basis in load management programs. In recent years, encouraged (or required) by their state regulators, an increasing number of utilities have issued requests for proposals for "negawatts" to be provided by CSP on a pay-for-performance basis. These efforts are often characterized as a move toward "outsourcing" provision of DR services, which in some cases are driven by the utility's need to meet aggressive demand-side reduction goals established by a state PUC. For example, California's investor-owned utilities (e.g., Southern California Edison, Pacific Gas and Electric, San Diego Gas& Electric) have signed long-term contracts with CSP in order to meet aggressive goals established by the California Public Utility Commissions (CPUC). As more utilities consider "outsourcing" DR programs, existing and new CSPs are now competing to provide this service and many CSPs now have dedicated "utility" sales staff to develop retail market leads by convincing utilities that CSPs can do it "cheaper, faster, and better."

CONCLUSION

This paper provides empirical evidence on the evolution of DR resources in U.S. electric power markets. This evidence shows that DR is a growing industry in the United States, as evidenced by the increasing number of entities that offer DR programs and dynamic pricing tariffs and the emergence of wholesale market DR programs. Based on data reported by utilities, ISOs and CSPs, the currently existing DR resource contribution, in terms of potential peak load reduction, has increased since 2006 by about 10%.

The vast majority of entities offering DR do so in the form of time-based retail rates; although, this type of DR accounts for a small share (<10%) of the total potential peak load reduction of all DR resources. The relative contribution of time-based retail rates among all DR resources is expected to increase over time as more utilities install interval meters for residential and small commercial customers that enable these types of rates as part of Advanced Metering Infrastructure (AMI) deployments.[15]

The existing DR resource potential ranges from 3 to 9% of a region's summer peak demand in most regions, with the notable exception of the Midwest Reliability Organization region where DR resources represent ~20% of summer peak demand.

With respect to assessing the accuracy of DR resources' expected performance, participants in energy market DR programs substantially overestimated their expected performance during declared program events, while participants in capacity market DR programs were much better at assessing their likely performance.

DR resources that participate in capacity markets typically face penalties for noncompliance, which is often not the case for DR resources that participate in wholesale energy markets. Thus program design (e.g. compensation levels, penalties for nonperformance, aggregation rules for small customers) can significantly influence the accuracy of DR resources' predicted performance.

There is significant year-to-year variability in DR performance at the portfolio level, particularly for economic DR programs. Over time, as customers gain experience and more ISOs (and utilities) offer economic DR programs, system operators will be in much better position to develop a "supply curve" that predicts the level of customer response over a range of different prices. This will be increasingly important if DR resources play a more substantial role in wholesale electricity markets.

However, the lack of standardized reporting practices and metrics for DR programs hinders reliability assessments. The North American Electric Reliability Council (NERC) has recognized this as a significant problem and has formed a Demand Response Data Task Force (DRDFT) to develop a system and protocol to collect DR event and market participation data to facilitate development of performance metrics [22].

Comprehensive surveys of utilities in recently formed organized markets (e.g. Midwest ISO and Southwest Pool Power) suggest that utilities are creating more flexible DR resources by adapting legacy load management and interruptible/curtailable DR programs to respond not only just to reliability concerns but also to reduce exposure to high market prices.

Finally, organized wholesale markets and policy support by the Federal Energy Regulatory Commission have facilitated new entry by curtailment service providers, which have expanded the DR industry and led to some product and service innovation.

REFERENCES

[1] U.S. Department of Energy. Benefits of demand response in electricity markets and recommendations for achieving them: Report to U.S. Congress pursuant to section 1252 of the Energy Policy Act of 2005. Washington D.C.: U.S. Department of Energy, 2006. See also http://eetd.lbl.gov/ea/EMP/reports/congress-1252d.pdf.

[2] Federal Energy Regulatory Commission. Assessment of demand response & advanced metering, staff report Docket Number AD-06-2-00. Washington, D.C.: Federal Energy Regulatory Commission, 2006. See also http://www.ferc.gov/legal/staff-reports/demand-response.pdf

[3] Woo CK, Kollman E, Orans R, Price S, Horii B. Now that California has AMI, what can the state do with it? Energy Policy 2008;36:1366-1374.

[4] Quantec, LLC, Summit Blue Consulting, Nexant, Inc.. Assessment of long-term system-wide potential for demand-side and other supplemental resources, report prepared for Pacificorp. Portland, OR: Pacificorp, 2007. See also http://www.pacificorp.com/ File/ File 75533.pdf.

[5] Federal Energy Regulatory Commission. Assessment of demand response and advanced metering, staff report. Washington, D.C.: Federal Energy Regulatory Commission, 2008. See also http://www.ferc.gov/legal/staff-reports/12-08-demand-response.pdf

[6] Hausman W, Neufeld J. Time-of-day pricing in the U.S. electric power industry at the turn of the century. The RAND Journal of Economics 1984; 15(1):116-126.

[7] Vickery W. Responsive pricing of public utility services. Bell Journal of Economics and Management Science 1971;2:337-346.

[8] Faruqui A, George S. Quantifying customer response to dynamic pricing. Electricity Journal 2005;18(4):53-63.

[9] Wellinghof J, Morenoff DL. Recognizing the importance of demand response: the second half of the wholesale electric market equation. Energy Law Journal 2007; 28(2):389-419.

[10] Barbose G, Goldman C, Neenan B. A survey of utility experience with real time pricing, report no. LBNL-54238. Berkeley, CA: Lawrence Berkeley National Laboratory, 2004. See also http://eetd.lbl.gov/EA/EMP/reports/54238.pdf.

[11] Barbose G, Goldman C, Bharvirkar R, Hopper N, Ting M, Neenan B. Real time pricing as a default or optional service for C&I customers: a comparative analysis of eight case studies, report no. LBNL-57661. Berkeley, CA: Lawrence Berkeley National Laboratory, 2005. See also http://repositories.cdlib.org/lbnl/LBNL-57661/

[12] Faruqui A, Sergici S. Household response to dynamic pricing of electricity: a survey of seventeen pricing experiments. Social Science Research Network Working Paper, 2008.

[13] Barbose G, Goldman C, Neenan B. The role of demand response in default service pricing, report no. LBNL-59737. Berkeley, CA: Lawrence Berkeley National Laboratory, 2006. See also http://eetd.lbl.gov/EA/EMP/reports/59737.pdf

[14] Neenan Associates, Lawrence Berkeley National Laboratory, Pacific Northwest National Laboratory. How and why customers respond to electricity price variability: a study of NYISO and NYSERDA 2002 PRL program performance, report no. LBNL-52209/PNNL#14220. Berkeley, CA: Lawrence Berkeley National Laboratory, 2003. See also http://eetd.lbl. gov/EA/EMP/reports/NYISO.pdf

[15] New York Independent System Operator. NYISO readies the grid for more wind: Press Release September 24, 2008. Albany, NY: New York Independent System Operator, 2008. See also http://www.nyiso.com/

public/webdocs/newsroom/press_releases/2008/NYISO_Readies_Grid_for_ More_Wind_09232008.pdf
[16] Bharvirkar R, Goldman C, Heffner G, Sedano R. Coordination of retail demand response with Midwest ISO wholesale markets, report no. LBNL-288E. Berkeley, CA: Lawrence Berkeley National Laboratory, 2008. See also http://eetd.lbl.gov/EA/EMP/reports/lbnl-288e.pdf
[17] Heffner G, Bharvirkar R, Goldman C. Retail demand response in Southwest Power Pool, report no. LBNL-1470E. Berkeley, CA: Lawrence Berkeley National Laboratory, 2009. See also http://eetd.lbl.gov/EA/EMP/reports/lbnl-1470e.pdf
[18] ISO New England . Introduction to demand response participation in New England's Forward Capacity Auction. Holyoke, MA: ISO New England Inc., 2007. See also www.isone.com/genrtion_resrcs/dr/broch_tools/intro_to_dr_in_fcm_training_21607.ppt
[19] ISO New England. Excel file: fca_monthly_obligation_including_zeros_final.xls. Holyoke, MA: ISO New England Inc. Downloaded May 27, 2008 from www.iso-ne.com, but no longer available.
[20] Federal Energy Regulatory Commission. Final rule order number 719, docket number RM07-19. Washington, D.C.: Federal Energy Regulatory Commission, 2008. See also http://elibrary.ferc.gov/idmws/common/opennat.asp?fileID=11833372
[21] PJM Interconnection, LLC. Proposed revisions to operating agreement: Members Committee meeting on January 22, 2009. Norristown, PA: PJM Interconnection, LLC, 2009. See also http://www.pjm.com/Media/committees-groups/committees/mc/20090122-item-02e-oatt-revisions-limiting-participation-in-dr.pdf
[22] North American Electric Reliability Corporation. Demand response data task force: preliminary report. Princeton, NJ: North American Electric Reliability Corporation, 2009. See also http://www.nerc.com/docs/pc/drdtf/DRDTF_Report_011708.pdf.

End Notes

[1] Other studies have developed alternative typologies to characterize demand response resources [3, 4] which are linked to program objective (e.g., system reliability or price response) or resource planning (e.g. firm vs non-firm resources) Given our subsequent focus on the two Federal Energy Regulatory Commission reports on demand response, we have chosen to use the typology found therein [2, 5].

[2] In estimating existing DR Resource contribution, FERC staff drew upon FERC survey responses and other sources (e.g. Energy Information Administration Form 861, Independent System Operator (ISO) or Regional Transmission Operator (RTO) DR program data).

[3] Prior to 2003, end-use customers had to enroll in both EDRP and SCR to receive both an up-front capacity payment and any energy payment that would be provided during program events.

[4] The methodology for triggering an RT-Price event has evolved over the past several years. At the program's inception, customers were able to curtail anytime between 7 a.m. and 6 p.m. if the day-ahead locational marginal price (LMP) or forecasts of real-time LMP exceeded $100/MWh. Starting in 2005, the event start time was scaled back to include only afternoon hours. More recently, the ISO has altered the trigger price to better track economic conditions in the wholesale market.

[5] The participation figures in 2007 for the Emergency program represent the enrollment in the "Energy Only" option, which is consistent with the voluntary nature of the Emergency program prior to that year. For reporting purposes, we have chosen to include all participants in the "Full" and "Capacity Only" options of the current Emergency program under the ALM category. This characterization is consistent with the historic ALM program, which included mandatory performance with the possibility for penalties and offered some form of a capacity payment.

[6] Enrollment in the NYISO DR program increased dramatically from 775 MW in 2001 (not shown in Fig. 2) to 2,025 MW in 2002.

[7] The subscribed load reductions prior to 2003 associated with participants jointly in EDRP and SCR were assigned to SCR, as that program has the threat of penalty for non-compliance.

[8] MWDRI was an initiative of the Organization of Midwest States (OMS) that resides primarily in the Midwest ISO (MISO) footprint.

[9] Four utilities were not members of MISO but operate in states that are part of OMS. Their responses were included in the study to provide a more comprehensive view of DR programs in the Midwest.

[10] Incentive-based DR programs were defined to encompass interruptible/curtailable rates, direct load control programs, and economic (e.g., demand-bidding, demand buy-back) programs. Time-based retail rates include real-time pricing and critical peak pricing. Finally, voluntary response programs were defined to represent any program where customers provided their "best-effort" to reduce consumption when requested but were not provided any compensation for doing so.

[11] Although five of the voluntary response programs had been called at least once, none had been evaluated at the time the survey was administered and thus respondents had no estimates of the programs' likely contribution to reducing peak loads.

[12] In Figs 5 and 6, the different sets (circles) in the Venn diagram represent the different dispatch conditions (i.e., system emergency, local reliability, or market price) and the indicated MW values represent the magnitude of committed load reductions from enrolled participants for the indicated set of conditions based on program administrators' estimates. Parts of the sets that overlap each other represent committed load reductions that can be dispatched for the different indicated dispatch conditions. For example, 20 MW of demand response in an SPP DLC program can be dispatched for any of the three conditions, whereas 39 MW can only be dispatched for local reliability reasons.

[13] ISO-NE did not reveal the name of entities that submitted offers in the FCA#1 in the public results; however project names were provided. Based on project names, which were often

descriptive enough to identify the submitting party, we were able to develop estimates of DR resources provided by a utility or CSP.

[14] PJM decided to put the onus on the enrolling customer's electric distribution company (EDC). The proposed tariff changes indicate once PJM receives a new customer registration, that customer's EDC will be notified and requested to submit within 10 days a copy of the relevant legislative or regulatory statute or decision expressly barring end-use customer participation [21]. This tariff language was approved by the PJM Members Committee on January 22, 2008 and will go to FERC for final approval and subsequent formal inclusion in PJM's Open Access Transmission Tariff (OATT).

[15] In the future, the relative contribution of dynamic pricing as a DR resource also depends on policy choices of state regulators (e.g. optional vs default tariffs), customer preferences and acceptance, marketing and education by utilities, and development and deployment of enabling technologies that facilitate price response.

In: Reduced Energy ...
Editor: Peter N. Ryan
ISBN: 978-1-62948-074-9

Chapter 3

MARKET AND POLICY BARRIERS FOR DEMAND RESPONSE PROVIDING ANCILLARY SERVICES IN U.S. MARKETS*

Peter Cappers, Jason MacDonald and Charles Goldman

FOREWORD

This report is one of a series stemming from the U.S. Department of Energy (DOE) Demand Response and Energy Storage Integration Study. This study is a multi-National Laboratory effort to assess the potential value of demand response and energy storage to electricity systems with different penetration levels of variable renewable resources and to improve our understanding of associated markets and institutions. This study was originated, sponsored, and managed jointly by the Office of Energy Efficiency and Renewable Energy and the Office of Electricity Delivery and Energy Reliability.

Grid modernization and technological advances are enabling resources, such as demand response and energy storage, to support a wider array of electric power system operations. Historically, thermal generators and hydropower in combination with transmission and distribution assets have been adequate to serve customer loads reliably and with sufficient power

---

* This is an edited, reformatted and augmented version of the Lawrence Berkeley National Laboratory, LBNL-6155E, dated March 2013.

quality, even as variable renewable generation like wind and solar power become a larger part of the national energy supply. While demand response and energy storage can serve as alternatives or complements to traditional power system assets in some applications, their values are not entirely clear. This study seeks to address the extent to which demand response and energy storage can provide cost-effective benefits to the grid and to highlight institutions and market rules that facilitate their use.

The project was initiated and informed by the results of two DOE workshops; one on energy storage and the other on demand response. The workshops were attended by members of the electric power industry, researchers, and policy makers; and the study design and goals reflect their contributions to the collective thinking of the project team. Additional information and the full series of reports can be found at www.eere.energy.gov/analysis/.

ACRONYMS AND ABBREVIATIONS

AMI	Advanced Metering Infrastructure
ARC	Aggregator of Retail Customers
AS	Ancillary Service
BA	Balancing Authority
CAISO	California Independent System Operator
DLC	Direct Load Control
DR	Demand Response
ERCOT	Electricity Reliability Council of Texas
FERC	Federal Energy Regulatory Commission
I/C	Interruptible and Curtailable
ILR	Interruptible Load for Reliability
ISO	Independent System Operator
ISO-NE	Independent System Operator New England
IRC	ISO/RTO Council
IRP	Integrated Resource Planning
IOU	Investor-Owned Utility
M$/yr	Million dollars per year
MCP	Market Clearing Price
MISO	Midwest Independent System Operator
MW-h	Megawatt per hour
NERC	North American Electric Reliability Corporation

NJBPU	New Jersey Board of Public Utilities
NPCC	Northeast Power Coordinating Council
PJM	PJM Interconnection, LLC
PSCo	Public Service of Colorado
RPS	Renewable Portfolio Standard
RTO	Regional Transmission Organization
SPP	Southwest Power Pool
U.S.	United States

EXECUTIVE SUMMARY

Introduction

The electricity grid requires various types of bulk power system services to maintain power quality, reliability, and security. Increasing penetration of renewable energy generation in U.S. electricity markets, driven primarily by state-level renewable portfolio standard (RPS) policies (Wiser et al. 2010), means that system operators will need to manage the variable and uncertain nature of many of these renewable resources to continue to meet their charter. This in turn is likely to require operational changes and procurement of greater quantities of various bulk power system services (NERC 2009). For example, system operators will likely need to procure, among other things, more ancillary services to fully accommodate the sizable addition of these variable generation resources (see Table ES-1).* Independent System Operators or Regional Transmission Organizations (ISO/RTO) in the U.S. typically procure various ancillary services via a centralized auction such that those who are committed to provide the services are paid a market- clearing price that fluctuates over time reflecting exigent system conditions, supply availability, and other factors. In jurisdictions without organized wholesale markets (hereafter referred to as non-ISO/RTO market environments), the Balancing Authority (BA) typically has a cost-based tariff, which is updated annually,

* In addition to ancillary services, many renewable integration studies (e.g., GE Energy 2008; Makarov et al. 2009; NERC 2009) have identified a need for greater ramping capability to track large but relatively slow changes in electricity production from variable renewable generation resource. Since this product has not yet been defined by NERC or any of the balancing authorities, we chose to focus on the bulk power system services that are currently defined.

that stipulates charges to transmission customers who do not self-supply or procure through a third-party sufficient capacity to meet their AS requirement.

Traditionally, ancillary services have been provided exclusively by generators. But over the past decade, alternative resources like demand response (DR) have become increasingly capable of providing such bulk power system services. Conceptual studies have argued that DR resources are well-suited to provide AS to the grid due in part to their fast response, distributed nature and the statistical reliability of large numbers of smaller resources (Kirby 2007; Callaway 2009; NERC 2009). These resources may be able to provide bulk power system services like AS at a lower cost and with a smaller carbon footprint than new conventional generation resources (Wellinghoff 2009), and can potentially be brought to market quicker as they don't have to go through lengthy permitting, siting and regulatory approval processes. Additionally, limited field tests of DR resources providing various forms of AS (Kirby and Kueck 2003; Todd et al. 2008; Kiliccote et al. 2009; Eto et al. 2012) have verified its technical capability.

However, while DR resources can technically provide these services, they may not do so until enabled by the entities and organizations that directly and indirectly affect a customer's interaction with the bulk power system. Federal regulators and reliability organizations create a framework for rules of operation through tariffs and other documents that affect the bulk power system of the various balancing authorities in America. However, federal regulatory influence to create opportunities for DR to more effectively participate as a resource is much greater in ISO/RTOs than for non-ISO/RTO BAs. Even if a balancing authority creates such opportunities, state regulators and legislators define the conditions under which electric utilities and aggregators of retail customers (ARC) can engage with customers.

As such, identification of barriers to DR resources' participation as an ancillary service provider and the entities responsible for addressing them is important at both the wholesale and retail level. Cappers, Mills et al. (2012), FERC (2009), and Kirby (2006) all identified various barriers limiting demand response resources from providing different bulk power system services (including ancillary services). FERC (2009) provided a high level summary of these issues and a more detailed assessment of barriers in California based on interviews. Kirby (2006) focused more on the technical requirements that DR resources must meet to provide ancillary services, while Cappers et al. (2012) identified the extent to which various DR opportunities could provide different bulk power system services based on current rate and program designs.

Table ES-1. Bulk power system operations affected by large-scale deployment of variable generation

	Time Scale				
Bulk Power System Operations	**Procurement or Schedule**	**Control Signal**	**Advance Notice of Deployment**	**Duration of Response**	**Frequency of Response**
Spinning Reserves (Contingency)	Days to hours ahead	<1 min	~1 min	~30 min	~20-200 times per year
Supplemental Reserves (Contingency)	Days to hours ahead	<10 min	~10-30 min	~Multiple hours	~20-200 times per year
Regulation (Normal Operation)	Days to hours ahead	~1 min to 10 min	None	< 10-min in one direction	Continuous

Adapted from Cappers et al. (2012).

Objective and Scope

In this study, we attempt to provide a comprehensive examination of various market and policy barriers to demand response providing ancillary services in both ISO/RTO and non-ISO/RTO regions, especially at the program provider level. It is useful to classify barriers in order to create a holistic understanding and identify parties that could be responsible for their removal. This study develops a typology of barriers focusing on smaller customers that must rely on a program provider (i.e., electric investor owned utility or IOU, ARC) to create an aggregated DR resource in order to bring ancillary services to the balancing authority.* The barriers were identified through examinations of regulatory structures, market environments, and product offerings; and discussions with industry stakeholders and regulators.

* Large customers (e.g., aluminum smelting) in most ISO/RTO environments can and often times do currently participate directly in the market as an ancillary service provider. As such, the barriers they faced bringing their capabilities to the bulk power system are somewhat different than those of smaller customers (e.g., retail office buildings) who must go through a program provider (i.e., IOU, ARC). For example, large customers usually have the requisite interval metering already installed, can provide load reductions that meet minimum size requirements, and can more readily afford to invest in the necessary telemetry requirements. As such, we are focusing in this study on the barriers standing in the way of BAs gaining access to smaller DR resources, as this is the group of customers that is still largely untapped. Many of the barriers listed here do still apply to larger electricity customers willing and able to go directly to the BA, but specifying where and explaining why this is or is not the case is beyond the scope of this paper.

In order to help illustrate the differences in barriers among various wholesale market designs and their constituent retail environments, four regions were chosen to use as case studies: Colorado, Texas, Wisconsin, and New Jersey. We highlight the experience in each area as it relates to the identified barriers.

Findings

Our review of the literature provided a useful starting point for our development of a typology. Since many of the specific barriers we identified could apply to several of the categories others' had constructed, we developed an alternative approach to characterizing barriers based on a logical progression from bulk power system service definitions to physical and financial requirements for DR resources to provide ancillary services through a program provider that better captures this interrelatedness (see Table ES-2).* Barriers associated with *Bulk Power System Service Definitions* relate to the way in which reliability organizations and BAs chose to define a service that includes/excludes certain classes of resources explicitly. These barriers must be dealt with first in order for a DR resource to even be able to provide these types of bulk power system services. Once a DR resource is eligible to provide such services, the rules developed by BAs that define the *Attributes of Performance* (e.g., minimum resource size) and the required *Enabling Infrastructure Investments* (e.g., automation and control technology, telemetry) for these services can create implicit barriers that may hinder program providers and DR resources from participating effectively in these AS markets. DR program providers and the participating customers must assess and decide whether the available revenues from participating in various AS

* A barrier type not explicitly mentioned in the typology presented in Table ES- 2 has to do with procedural issues. Procedural barriers generally relate to the process by which other barriers are removed. As such, these are not direct barriers to entry but rather introduce additional transaction costs, both in time and financial resources, in implementing solutions to any identified barriers. For instance, ISO/RTOs that are attempting to change their product definitions must: initiate internal and external stakeholder processes that require among other things overcoming preconceptions about capabilities of particular resources; seek approval from regulators before any changes may be made to operating practices; and then contend with both financial and human capital resource constraints that can significantly slow market changes. A similar issue is found with retail electric utilities who must address certain barriers through internal and external stakeholder processes that include seeking regulatory approval for tariff changes, which likewise can be a time-consuming process. These less direct barriers permeate the entire structure of the typology presented in Table ES- 2.

markets are sufficient (*Revenue Availability*) and can be captured with enough certainty (*Revenue Capture*) to meet simple payback periods for customers or return on investment hurdle rates of IOUs or ARCs for all fixed and variable enabling infrastructure investment costs. The regulatory compact between utility and regulator, along with other statutes and decisions by state policymakers, may also create barriers that *Program Providers* must overcome in order to pursue DR resources more vigorously as an ancillary service provider.

One of the most effective ways to remove these barriers is to alter the requirements imposed on DR resources wishing to provide AS (see Table ES-3). By acknowledging that demand response is fundamentally different than a generator, many ISO/RTOs are currently finding ways to alter the requirements to provide these services such that the quality of the service they are procuring is maintained but the pool of resources that can provide it is expanded. In contrast, most non- ISO/RTO BAs do not appear to be as far along in their attempts to better integrate non-traditional resources as ancillary service providers. Additionally, with advancements in technology through research and development efforts and with increased market adoption, the cost of automation and control technology and other forms of enabling infrastructure investments should continue to drop making participation as an AS provider more cost effective. Increases in benefits, through market rule changes (e.g., scarcity pricing, reserve demand curves) can likewise contribute to an increase in the cost effective procurement of AS from demand response. Finally, altering the process by which program providers do business (e.g., changes in the utility's business model) should help further facilitate increased interest in pursuing DR resources as a viable AS provider.

The four regions focused on in this study illustrate these various barriers and ways in which the regions have or are attempting to address them.

Colorado has a vertically integrated retail utility environment that operates within a non- ISO/RTO BA footprint. This results in little to no opportunities for ARCs directly soliciting customers to bring DR resources forward to provide AS unless there is someone on the other end to buy their services. The BA doesn't provide such a "marketplace", so the ARCs must develop relationships with the utilities in the BA's jurisdiction in order to play some sort of role. The investor-owned utilities, as the sole provider of electricity service to customers, have only modest profit motives to pursue non-generation resources, like DR, as an AS provider. Colorado utilities and state regulators have not placed expansion of DR programs to provide ancillary services as a high priority given the excess capacity situation currently experienced by the state's utilities. If the capacity situation tightens and/or the distribution system requires attention due to increased penetration of distributed variable renewable resources, which may occur later this decade due to the state's RPS requirements, state regulators as well as the utilities could be more inclined to consider such new DR opportunities.

Table ES-2. Applicable entities and organizations responsible and affected by barriers

	Reliability Council	BA	IOU	ARC	Utility Regulator	End-use Customer
Bulk Power System Service Definitions	*	*,●	●	●		
Attributes of Performance		*	●	●		●
Enabling Infrastructure Investments		*	●	●		●
Revenue Availability		*	●	●		●
Revenue Capture		*	●	●		●
Program Providers		*	●	●	*	●

*- Entity/Organization responsible for creating the barrier.

● - Entity/Organization affected by the barrier.

Table ES-3. Actions required to overcome barriers

	Change Definition	Change Requirement	Change Process	Reduce Costs	Increase Benefits
Bulk Power System Service Definitions	♦				
Attributes of Performance		♦			
Enabling Infrastructure Investments		♦			
Revenue Availability					♦
Revenue Capture		♦			
Program Providers			♦		

♦ - Primary action to overcome barrier.

- Secondary action to overcome barrier.

Texas, with its open retail market and integrated wholesale market, appears to have conditions that are highly conducive for DR resources to readily provide ancillary services. ERCOT has indeed attracted a substantial amount of demand response to its markets, enrolling ~2,400MW of capacity to provide spinning reserves (Patterson 2011). The vast majority of this is provided by industrial facilities who have peak demands of 10 MW or more and utilize under-frequency relays that were installed through utility-sponsored instantaneous interruptible tariffs years prior to the advent of the organized wholesale market (Zarnikau 2010). While ERCOT has been very successful at operating with sizable amounts of ancillary services provided primarily by these large customers (e.g., DR resources can and often do provide up to 50% of the spinning reserve requirement), relatively few new DR resources have been brought to market that aggregate smaller customers. Conversations with program providers and regulatory staffers illustrate the challenges that the open retail market creates when trying to attract these smaller DR resources. The distribution utilities, who no longer have supply obligations to customers, have little profit motive to pursue such programs, which means regulators must get involved to either create the business case or mandate that programs be offered. Competitive retail electricity suppliers do not yet readily see a value proposition in offering the requisite types of enabling technology as an additional service due in part to short customer contract lengths which do not exceed the return on investment hurdle rates on the equipment. The current wholesale "energy-only" market design does not provide a reservation payment that guarantees a longer term (e.g., quarterly, semi-annual, or annual) revenue stream compared to other ISO/RTOs that administer capacity markets. This adds risk into the decision to invest in enabling infrastructure at customer premises. Reserve margins that exceeded ERCOT requirements by 2-4 percentage points coupled with major changes at ERCOT to a nodal market over the last several years resulted in time and effort not being directed towards creating greater opportunities for DR resources to participate in the market. Although with more recent reserve shortages, ERCOT and the Texas Public Utility Commission are jointly working through the stakeholder process to resolve these issues and improve the environment for DR in general (see Public Utility Commission of Texas Project No. 40000 for more details).

New Jersey is in a similar situation as Texas but for somewhat different reasons. The incumbent utilities' business model creates little incentive for them to pursue new DR resource in any fashion, let alone as an ancillary service. New Jersey allows retail competition and utilities currently provide electric commodity and ancillary services as a bundled product for default service customers through a multi-year contracting (i.e., auction) process. The costs of this bundled product are completely passed through to these customers, which provides little incentive for the utility to pursue lower cost options to provide these services *outside of the auction process* nor any ability to do so within the auction process since the products are bundled. So even though PJM is on the forefront of creating and expanding opportunities for DR resources to provide various forms of ancillary services, it is only ARCs or competitive retail providers going directly to the market or through bilateral contracts with third-party suppliers[*] who are likely to bring resources to market, absent state regulatory engagement.[#]

Wisconsin, with its regulated retail environment and restriction on ARCs providing DR programs, faces both regulatory and utility business model barriers that limit their utilities' interest in pursuing DR resources as an ancillary service provider. Electric utilities in this state are vertically integrated. Since they rely on their own generation assets to serve customers' needs, any reductions in non-fuel operating expenses from more efficiently operating this fleet of resources can be captured by the utility but only until new rates are set which reflect these now lower costs. Although these types of DR programs also create an opportunity for the state's utilities to convert reserved generation capacity into energy which can be sold off-system, state regulators do not allow the utility to retain any of these profits but instead require the utility to turn them over to ratepayers. Furthermore, the electric utilities profit from investing in capital assets, like new generating stations, which is often not true for DR investments. As such, Wisconsin utilities have only modest financial incentives to pursue DR resources in general, while state regulators have restricted ARCs from doing business in the state. Should state regulators choose to alter the utility's business model making the pursuit of DR resources as an AS provider more lucrative, the utility must alter its existing tariff to expand the conditions under which current and future DR resources can be dispatched. Regulators could also choose to lift their

[*] Third-party electricity suppliers provide capacity, energy, and ancillary services as a portfolio to competitive electricity suppliers or the incumbent utility for default service. The goal of third-party suppliers is to provide the product at least-cost, which may entail pursuing DR as a resource within the portfolio.

[#] The New Jersey Board of Public Utilities (NJBPU) in 2008 did get involved by ordering the utilities to augment DR program payments for new resources participating in PJM's 2009 Interruptible Load for Reliability (ILR) program through a series of one-year pilots with a goal of fostering cooperation with ARCs (NJBPU 2008). Two years later, the NJBPU stopped all utility pilot programs due to a lack of funding; leaving ARCs to continue enrolling customers directly in PJM programs without any augmented payment (NJBPU 2010).

moratorium on ARCs to further facilitate access to DR resources that can provide ancillary services. However, at this time, based on our interviews, the Commission does not appear inclined to look into these issues in the near future.

These four regions included as case studies in this report are likely representative of the experience in similar wholesale and retail market environments across the country. Our assessment of the barriers to smaller DR resources providing ancillary services illustrates that the issues span nearly all entities and organizations in the chain that connects wholesale markets to retail customers. Fundamentally, in a highly regulated system, incentives must be aligned among the balancing authority, utilities, and program providers in order for DR resources to reach the system operator, where AS is procured. Most of these entities have little to no incentive to address these barriers on their own, even if society as a whole could greatly benefit from the effort. Thus, many different but disparate parties will need to work together for the common good in order for smaller DR resources to reach their full potential as a provider of ancillary services.

1. Introduction

The electricity grid requires various types of bulk power system services to maintain power quality, reliability, and security. Increasing penetration of renewable energy generation in U.S. electricity markets, driven primarily by state-level renewable portfolio standard (RPS) policies (Wiser et al. 2010), means that system operators will need to manage the variable and uncertain nature of many of these renewable resources to continue to meet its charter. This in turn is likely to require operational changes and procurement of greater quantities of various bulk power system services (NERC 2009). For example, system operators will likely need to procure, among other things, more ancillary services to fully accommodate the sizable addition of these variable generation resources (see Table 1).[1] The type and level of additional bulk power system services needed depend on a variety of factors from location to adoption of different technologies. In California, for example, an increase in the amount of regulation capacity is expected to largely accommodate rapid changes in solar power plant output (Makarov et al. 2009; Helman 2010), while in Texas, studies suggest that the need for spinning reserves (called responsive reserves in ERCOT) is growing to accommodate large unexpected wind ramps (GE Energy 2008).

Table 1. Bulk power system operations affected by large-scale deployment of variable generation

	Time Scale				
Bulk Power System Operations	**Procurement or Schedule**	**Control Signal**	**Advance Notice of Deployment**	**Duration of Response**	**Frequency of Response**
Spinning Reserves (Contingency)	Days to hours ahead	<1 min	~1 min	~30 min	~20-200 times per year
Supplemental Reserves (Contingency)	Days to hours ahead	<10 min	~10-30 min	~Multiple hours	~20-200 times per year
Regulation Reserves (Normal Operation)	Days to hours ahead	~1 min to 10 min	None	< 10-min in one direction	Continuous

Adapted from Cappers, Mills et al. (2012).

Currently these various forms of ancillary services (e.g., reserves and regulation) are products which Balancing Authorities (BA) are responsible for procurring. Where the BA runs an organized wholesale market in the U.S., commonly referred to as an Independent System Operator or Regional Transmission Organization (ISO/RTO), ancillary services are typically procured via a centralized auction such that those who are committed to provide the services are paid a market-clearing price. In jurisdictions that have no organized wholesale markets, the BA typically has a cost-based tariff that stipulates charges to transmission customers who do not self- supply or procure through a third-party sufficient capacity to meet their AS requirement (hereafter referred to as non-ISO/RTO BA environments).

Traditionally, ancillary services have been provided exclusively by generators. However, over the past two decades, alternative resources like demand response (DR) have become increasingly capable of providing such bulk power system services. Conceptual studies have argued that certain forms of DR resources are well-suited to provide AS to the grid due in part to their fast response, distributed nature and the statistical reliability of large numbers of smaller resources (e.g., Kirby 2007; Callaway 2009; NERC 2009). These resources may be able to provide bulk power system services like AS at a lower cost and with a smaller carbon footprint than new conventional generation resources (Wellinghoff 2009), and can potentially be brought to market quicker as they do not have to go through lengthy permitting, siting and regulatory approval processes. Additionally, limited field tests of DR resources that provided various forms of AS (Kirby and Kueck 2003; Todd et al. 2008; Kiliccote et al. 2009; Eto et al. 2012) have verified their technical capability.

However, while DR resources can technically provide these services, they may not do so until enabled by the entities and organizations that directly and indirectly affect a customer's interaction with the bulk power system. Federal regulators and reliability organizations create a framework for rules of operation through tariffs and other documents that affect the bulk power system of the various balancing authorities in America. However, federal regulatory influence to create opportunities for DR to more effectively participate as a resource is much greater in ISO/RTOs than for non-ISO/RTO BAs. Even if a balancing authority creates such opportunities, state regulators and legislators define the conditions under which electric utilities and aggregators of retail customers (ARC) can engage with customers.

As such, identification of barriers to DR resources' participation as an ancillary service provider and the entities responsible for addressing them is important at both the wholesale and retail level. For example, many of the current BA ancillary services procurement rules and qualification policies were initially designed under a "generator-only" paradigm. As such, the playing field for provision of such ancillary services has not been level, generally favoring traditional generation resources for which the services were originally designed. The Federal Energy Regulatory Commission (FERC) issued several recent decisions that attempt to address and rectify the situation (FERC 2008, 2011a) in ISO/RTO jurisdictions and mandate that fast acting resources be compensated for the additional security that their flexibility brings the grid (FERC 2011b).[2] Furthermore, utility demand response programs, which have historically been the mainstay for customer DR opportunities, have not traditionally provided ancillary services. There are, however, non-utility DR program providers (i.e., ARCs) whose business model is predicated on enabling and subscribing customers to provide these types of services; yet they may be precluded from doing so based on state statutes or regulatory decisions. Therefore, state regulators and legislators have a role to play in creating opportunities for retail entities (electric utilities and ARCs) to see value in altering or expanding their program offerings.

Although some of the barriers hindering DR resources from providing ancillary services have been identified and addressed (e.g., FERC), a comprehensive assessment of these barriers could help the electric industry identify the entities most capable of undertaking the necessary actions to overcome them such that DR resources can reach their full potential. Cappers et al. (2012), FERC (2009), and Kirby (2006) all identified various barriers limiting demand response resources from providing different bulk power system services (including ancillary services). FERC (2009) provided a high level summary of these issues and a more detailed assessment of barriers in California based on interviews. Kirby (2006) focused more on the technical requirements that DR resources must meet to provide ancillary services, while Cappers et al. (2012) identified the extent to which various DR opportunities could provide different bulk power system services based on current rate and program designs.

In this study, we attempt to provide a comprehensive examination of various market and policy barriers to demand response resources associated with smaller customers who must go through a program provider in order to provide ancillary services in either an ISO/RTO or non-ISO/RTO region.[3] This report is organized as follows. We first present a general description of U.S. electricity market environments and include a proposed typology for the assessment of barriers to DR resources providing ancillary services (see sections two and three). That framework is used to organize the subsequent discussion in sections four through nine of individual barriers (e.g., barriers that arise because of bulk power system service definitions, enabling infrastructure barriers, and barriers to participation by certain types of DR program providers). In order to help illustrate the differences among wholesale markets and their constituent retail environments, four states were chosen to use as case studies: Colorado, Texas, Wisconsin, and New Jersey. We highlight the experience in each area as it relates to the identified barriers throughout the paper and then summarize that experience in the conclusions (see section 10).

2. Influential Entities and Organizations in Wholesale/Retail Market Environments

A variety of different entities and organizations directly and indirectly affect an individual customer's ability to provide a bulk power system service to the balancing authority (see Figure 1). Retail entities, specifically electric utilities and aggregators of retail customers, harness the demand response capabilities of electric customers through tariff rates and DR programs that are approved or allowed by the applicable state or local regulatory authority. Although the balancing of supply and demand is managed at the wholesale level by a balancing authority, this entity functions under open access transmission tariffs approved by the Federal Energy Regulatory Commission (FERC) and must operate their system in accordance with rules set by regional reliability organizations. These rules are derived from enforceable standards set by the North American Electric Reliability Corporation (NERC). Thus, the role that DR resources can play in providing ancillary services (or any bulk power system service) is contingent on what opportunities state and local regulators are willing to promote and approve, as well as how customers' response to the DR opportunities can be integrated into the bulk power system both physically (due to reliability rules) and financially (due to tariffs and market rules). The barriers that must be addressed by these various entities and organizations differ depending upon the type of wholesale and retail electricity market environment that exists.

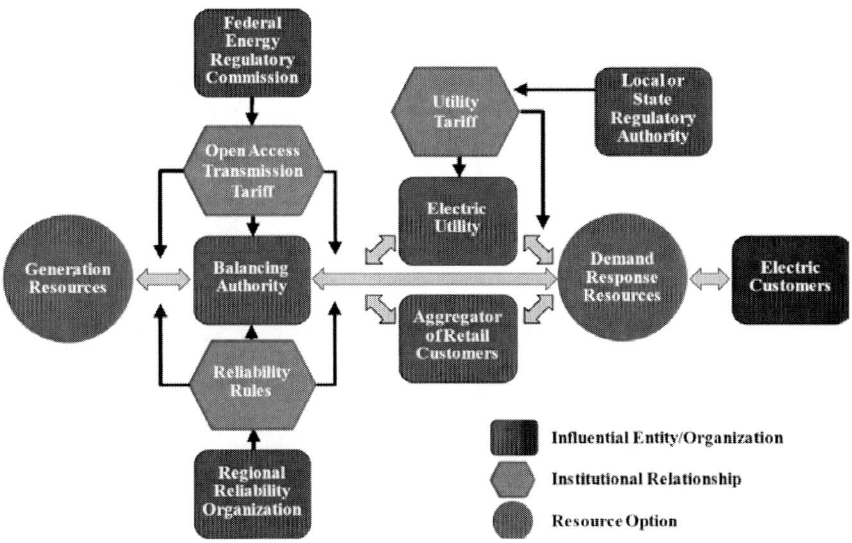

Adapted from Cappers et al. (2012).

Figure 1. Entities and organizations that influence relationships between resources and the bulk power system.

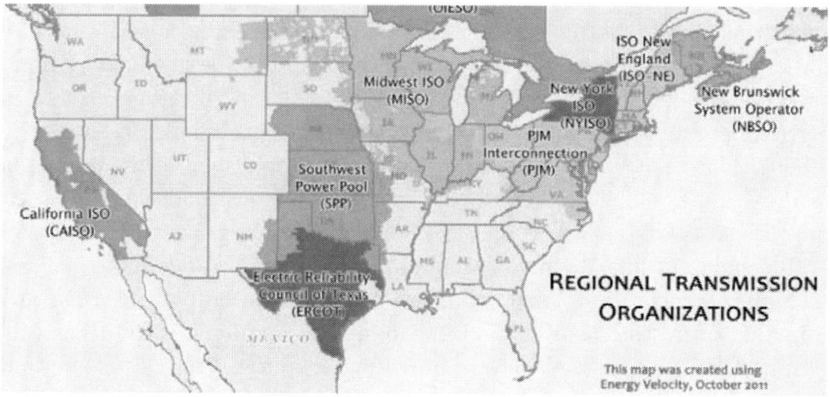

Source: FERC (2012).

Figure 2. Map of ISO/RTO balancing areas in the U.S.

There are seven organized wholesale electricity markets in the U.S. operated by BAs known as Independent System Operators or Regional Transmission Organizations (see Figure 2). Each ISO/RTO has its own set of rules that dictate the structure of its various markets for different bulk power system services, but all are based on a basic auction concept that brings multiple buyers and sellers together. These markets provide transparent price signals for all transacted products, including ancillary services, which are not present in BAs solely with bilateral markets (the remainder of the country). ISO/RTOs can be characterized by the types of wholesale markets they run and the way in which they operate. Some ISO/RTOs have a market design that includes a forward capacity market which provides a reservation payment to all resources selected to provide planning reserves (e.g., generators, demand response), while others just have an "energy-only" market design where no such capacity market exists and thus no upfront revenue source is available. In contrast, all ISO/RTOs procure or plan to procure ancillary services and energy in day-ahead forward and/or real-time spot markets. Energy and ancillary service markets are co-optimized in all ISO/RTOs but one to jointly procure the necessary levels of ancillary services and energy at least cost. The remaining ISO/RTO (i.e., SPP) plans to operate AS markets and co-optimize procurement by 2014. ISO/RTOs also differ in the way they determine their required ancillary service capacity to procure. Some ISO/RTOs set a flat amount that must be procured for every hour of the year (e.g., ERCOT); others determine a monthly requirement that varies every hour (e.g., MISO); and still others establish their requirement as a function of the daily forecast of system load (e.g., PJM).

In regions without ISO/RTOs (see Figure 2), the BA runs a day-ahead unit commitment model and in-day dispatch model of resources for energy and ancillary services like their ISO/RTO counterparts; however, these are predominantly based on schedules provided by vertically- integrated utilities and the non-utility generators producing the energy or providing the ancillary services and usually prices are not produced out of these operating platforms. The utilities will generally notify the BA that they have a balanced schedule for all necessary services, which they will have forecasted and subsequently procured from a portfolio of their own fleet of resources, long-term power purchase agreements with other resources, and short-term bilateral contracts with other resources. In contrast to the ISO/RTO markets, any shortfalls in meeting this balance is charged at the transmission tariffs' stipulated value for providing these services based on a FERC-approved cost-of-service analysis. Although power exchanges do enable trading of electric commodities in these regions, these markets are voluntary and simply match one buyer and one seller.

As with wholesale electric markets, there are also significant differences in the structure and design of retail electricity markets in the United States. First and foremost, electric utilities can differ in the way they are owned and governed. Electric cooperatives are generally owned by their members, who also happen to be ratepayers, and governed by a board of directors. Public utilities are owned and generally governed by the municipalities they serve. Private shareholders own investor-owned utilities, which are regulated by either an elected or state-appointed group of utility commissioners. For purposes of this report, we will be focusing exclusively on IOUs when determining the market and policy barriers hindering DR resources from providing ancillary services, as the business model and regulatory issues for municipalities and rural cooperatives can be quite different. Among investor owned utilities, some states allow retail competition where the utility retains monopoly status for the provision of transmission and distribution services but not electric supply service. Customers, therefore, are free to choose their electricity supplier, even though in all of these states, except Texas, utility commissions have required the monopoly transmission and distribution utility to serve as the default electricity supplier. In states that do not allow retail competition, IOUs maintain their monopoly status as the provider of all three services. Electric utilities can also be characterized by their ability to own and operate generation assets. A vertically integrated utility has its own fleet of generation resources that are used to serve the electricity needs of its customers, whereas a distribution utility may have divested or sold some or all of these generation assets at some point in the past. Such distribution utilities must now contract with non-utility electricity suppliers (e.g., private power generators) to procure the necessary energy, capacity and ancillary services to meet its customers' demands through long-term contracts and short-term purchases on the spot market. In addition, some investor-owned utilities, but not all, have the ability to retain some or all of the profit associated with either selling excess power they have produced and/or arbitraging power purchased through long-term forward contracts in the spot electricity market. Lastly, the IOU has, up until about the turn of the millennium, been the sole provider of demand response programs. In some jurisdictions, third party aggregators of retail customers are now able to offer DR programs that compete directly with the utility's offered programs.[4]

In this study, we selected four distinct states to represent the diversity of wholesale and retail environments in the United States: Colorado, New Jersey, Texas, and Wisconsin (see Table 2). Colorado is the lone state in our analysis that is in a non-ISO/RTO BA environment. Electric utilities in Colorado are vertically integrated and do not face retail electricity supply competition. Wisconsin electric utilities are likewise vertically integrated and have no competition for supply service, but operate within the MISO footprint. Texas and New Jersey both have retail competition, where customers can procure electric supply from any number of different entities within the ERCOT and PJM footprints, respectively. The incumbent electric utilities in New Jersey must act as the default service provider for all customers in their service territory, whereas the incumbent electric utilities in Texas have no such supply obligation.

3. Market and Policy Barrier Typology

When considering the market and policy barriers to provision of ancillary services by demand response resources, it is useful to classify barriers in order to create a holistic understanding and identify parties that could be responsible for their removal. This study develops a typology for barriers based on examinations of regulatory structures, market environments, and product offerings and conversations with industry stakeholders and regulatory staff.

Based on our literature review, we identified several sources that identified and characterized a typology of DR resource barriers. For example,

FERC (2009) categorized barriers into four groups: regulatory, economic, technological and other. Regulatory barriers are those that are created by entities and organizations that operate, regulate and/or dictate the definitions and requirements imposed on providers of AS. Economic barriers are those that affect the financial willingness of entities and organizations to pursue opportunities to provide AS. Technological barriers to implementation of demand response are those that require a customer and/or program provider to invest in new control, automation, metering, and/or communications equipment. Other barriers simply fall outside of these three categories.

Our review of the literature provided a useful starting point for our development of a typology. Since many of the specific barriers we identified could apply to several of the categories others' had constructed, we developed an alternative approach to characterizing barriers that captures their interrelatedness (see Figure 3), from the perspective of DR program providers. Barriers associated with *Bulk Power System Service Definitions* relate to the way in which reliability organizations and the BAs chose to define a service that includes/excludes certain classes of resources explicitly. These barriers must be dealt with first in order for a DR resource to even be able to provide these types of bulk power system services.

Once a DR resource is eligible to provide such services, the rules developed by balancing authorities to define the *Attributes of Performance* (e.g., minimum resource size) and the required *Enabling Infrastructure* (e.g., telemetry requirements) for these services can create implicit barriers that may hinder program providers and DR resources from participating effectively in these AS markets. Such requirements may limit and constrain interest in participation if these requirements are perceived to be too onerous or costly. DR program providers and their participants must assess and decide whether the available revenues from participating in various AS markets are sufficient (*Revenue Availability*) and can be captured with enough certainty (*Revenue Capture*) to meet simple payback periods of customers or return on investment criteria of IOUs and ARCs for all fixed and variable enabling infrastructure investment costs.[5] *Program Providers* must also specifically contend with a host of additional issues (e.g., viability of business model, ability to offer programs) that may limit their interest in pursuing DR resources as an ancillary service provider. Each of these barrier categories is discussed in more detail below and in the appendices.

Table 2. Wholesale and retail difference among case study regions

		Colorado	Texas	Wisconsin	New Jersey
Wholesale	Dominant Balancing Authority	PSCo	ERCOT (ISO)	MISO	PJM (RTO)
	Capacity Market Exists?	N/A	No	Yes*	Yes
	Co-optimized Energy & Ancillary Services?	N/A	Yes	Yes	Yes
	AS Capacity Determination?	Constant, Annually	Constant, Annually	Daily shape, monthly	Based on Forecasted Load
Retail	Allows retail competition?	No	Yes	No	Yes
	Distribution IOU?	No	Yes	No	Yes

* MISO's capacity market is strictly voluntary and covers capacity obligations only one month ahead.

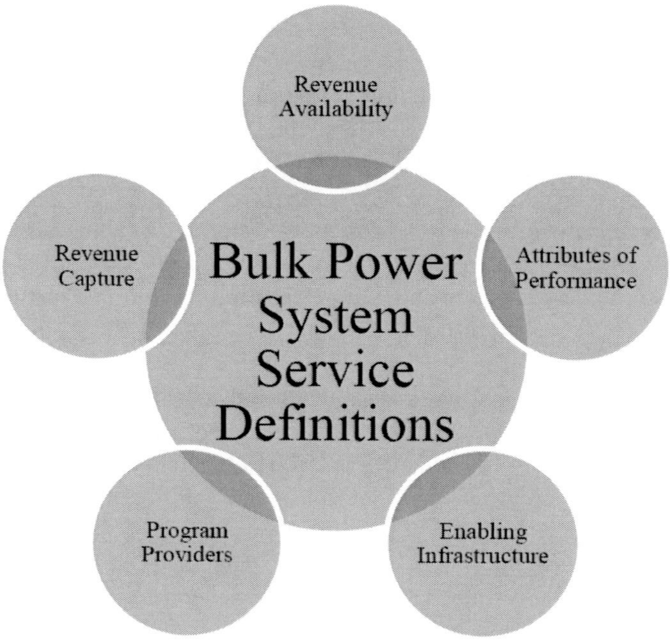

Figure 3. Conceptual framework for a typology of barriers to demand response resources providing ancillary services.

A final barrier type not explicitly mentioned in the typology presented in Figure 3 has to do with procedural issues. Procedural barriers generally relate to the process by which other barriers are removed. As such, these are not direct barriers to entry but rather introduce additional transaction costs, both in time and financial resources, in implementing solutions to any identified barriers. For instance, ISO/RTOs that are attempting to change their bulk power system service definitions must: initiate internal and external stakeholder processes that require overcoming preconceptions about the capabilities of particular resources; seek approval from federal regulators before any changes may be made to operating practices; and then contend with both financial and human capital resource constraints that can significantly slow market changes. A similar issue is found with retail electric utilities who must address certain barriers through internal and external stakeholder processes that include seeking regulatory approval for tariff changes, which likewise can be a time-consuming process. These more indirect barriers permeate the entire structure of the typology presented in Figure 3.

4. Bulk Power System Service Definitions Barriers

Barriers associated with bulk power system service definitions are those that preclude demand response from participating as an ancillary service resource by defining specific resource types eligible to provide a particular product that do not include or effectively exclude demand response. These definitional barriers arise in both North American standards and regional product definitions.

Regional reliability organizations define the bulk system reliability requirements, including ancillary services, for the areas under their jurisdiction. In defining the service, such entities may explicitly identify which type of resource can provide that service. The North American Electricity Reliability Corporation, a reliability standards setting institution, has defined all ancillary services such that demand response is included among the resources that may provide the services (NERC 2010). However, some regional reliability organizations have yet to adopt similar standard language and do not allow provision of some ancillary services from DR resources. As an example, the Western Electricity Coordinating Council, which covers Colorado, does not currently allow DR resources to provide regulation service

(WECC 2007); however, all three of the reliability councils covering the ISO/RTOs included in this study (i.e., ERCOT, MISO and PJM) have adopted language similar to NERC's standard.

A balancing authority also has an opportunity to further refine the requirements for the types of resources that can provide its reliability services as well as those for which it has integrated into its market operations, to the degree such markets exist. The three ISO/RTOs in the case study regions have adopted ancillary service definitions that allow provision with demand response directly in their markets, but this is not true of some other ISO/RTOs in the US. For example, the Northeast Power Coordinating Council (NPCC) complies with NERC ancillary service definition standards, but ISO New England (ISO-NE), an ISO/RTO within NPCC's jurisdiction, has not yet altered its market design to include DR resources in regulation markets. However, ISO-NE is testing these resources' ability to provide the services through an "out-of-market" pilot program (ISO-NE 2012). The open access transmission tariff filed with FERC by Public Service Company of Colorado (PSCo), the dominant BA in Colorado, identifies the various types of resources that may provide the different types of ancillary services (FERC 2010). Non- generation resources are specifically identified for each of the ancillary services as eligible to provide the service, with Supplemental Reserves explicitly identifying interruptible loads and pumped hydro storage as eligible resources. In spite of this, non-generation resources are unable to offer regulation and spinning reserves to PSCo because WECC has precluded such resources in its service definitions. Once WECC alters its service definitions, PSCo would be able to begin accepting non-generation resources as spinning reserves and regulation providers.

5. Attributes of Performance Barriers

Even if DR is eligible to provide various forms of AS, DR resources and their program providers may still have to overcome a number of barriers that originate at the BA due to its requirements to be an ancillary service provider. AS products and their associated qualification requirements were designed when the electricity system was made up primarily of large generators. Operating models for resources were fairly homogenous due to the similarity of operational characteristics of the fleet of generators. This led to rules that did not necessarily reflect the technical requirements of a service, but rather the focus was often on the technical capabilities of the existing resources that could provide that service. Given the stark differences in characteristics associated with other types of resource (e.g., demand response, battery storage) that were not considered when these operating models were designed and built, such resources have found it difficult to readily integrate themselves into the existing operating model, let alone capture any additional value that non-traditional but more flexible alternative resources could provide.

Demand response is a fundamentally different resource than a traditional generator. Individual DR resources vary substantially in size. In particular, they can be very small; far smaller than even the smallest of generators. The DR resource can be distributed behind several retail electricity meters, and often measured as part of a whole system of loads, some of which can and cannot be controlled for purposes of providing a bulk power system service. While it makes sense to require fast, continuous monitoring of each large generator that is providing a large amount of an ancillary service so that the system operator is immediately aware if the generator stops responding, this same requirement may not be appropriate for each individual load. The failure of an individual load to respond does not have the same reliability consequences and statistical methods for ensuring (and verifying) response of the DR resource in aggregate may be more appropriate (Kirby 2006). Ancillary services response and monitoring requirements should be based on the technical reliability requirements of the power system and not be defined based on the characteristics of the historic AS providers.

In spite of these differences, DR may be better suited than a generator to provide certain bulk power system services under certain circumstances. DR resources vary in the speed with which they can provide various bulk power system services; some can be considerably faster than traditional generators when automated (Eto et al. 2012).

Many DR resources are more resilient to rapid changes in electricity consumption than traditional generators are to rapid changes in electricity production (i.e., cycling).

All of these differences make adapting the current operating procedures and market rules, which were developed largely for thermal generators, to alternative resource types difficult, but important in order to potentially gain access to these types of resource, and as such may require changes to those rules in order to address barriers to their participation.

In defining the performance attributes required to provide certain bulk power system services in ISO/RTO jurisdictions (ERCOT 2012a, b, c; IRC 2012; MISO 2012; PJM 2012a, b), ISO/RTOs have included rules and requirements that may limit the pool of eligible demand response resources to provide AS (e.g., limitations on resource size, the ability to aggregate multiple small resources, geographic boundaries of aggregation, and symmetric response capabilities).[6] Traditional generators are relatively large, (e.g., 100–2,500 MW) which resulted in bulk power system unit commitment and dispatch computer platforms being designed to deal with large units of power (i.e., whole or tenths of MWs). The resulting service requirements, therefore, largely relied on a common minimum resource size of 1 MW, but only the largest and most sophisticated individual customers are capable of providing a bulk power system service at so large a level. Recently, some ISOs (e.g., PJM and ERCOT) have begun to relax this size requirement (e.g., reducing minimum resource size to 0.1 MW) to promote entry of new technologies.

DR program providers may also reach the minimum resource size requirement by aggregating many smaller, independently metered customers as a single DR resource portfolio. This is not always allowed in ISO/RTO markets, but some ISO/RTOs are implementing pilot projects to test viability in their systems (e.g., ISO-NE). Geographic boundaries exist in wholesale power markets which are based on physical limitations that have implications for the required procurement of certain bulk power system services (e.g., capacity, spinning reserves) and their settlement. ISO/RTOs, therefore, set rules on the geographic boundaries of aggregations of DR resources which may limit the ability of a program provider to meet the minimum resource size requirement. For regulation, a requirement for resources to provide equal capacity to move in both directions (i.e., reduce consumption and increase consumption) also effectively restricts the size of the pool of DR resources. This is due to the fact that many demand response resources may be more flexible in one direction than another (e.g., shedding load vs. increasing load).

These types of rules impact the program provider's ability to make DR resources sizeable enough to enter the market and can severely limit aggregate participation, unless the ISO/RTO broadens the rules and provides mechanisms in their backend software to accommodate smaller and/or aggregated resources.

These same types of issues also need to be addressed in non-ISO/RTO environments for DR (and other alternative flexible resources) to be considered on par with generation resources.

Table 3 indicates the status of some of these barriers in the case study regions and illustrates the degree to which different jurisdictions have addressed the previously cited barriers. PJM has the most favorable rules for demand response participation, given its low minimum size threshold and ability for ARCs and utilities to readily aggregate DR resources. ERCOT also has favorable market rules due to the ability to asymmetrically bid regulation capacity and its low minimum resource size. MISO has not gone as far as PJM and ERCOT in revising its market rules to facilitate entry by DR resources to provide ancillary services. At present, PSCo, the BA in Colorado, does not appear to have begun to deal with these issues.

6. Enabling Infrastructure Investment Barriers

Current DR programs are typically designed to elicit load reductions during periods of high system demand, when system reliability is threatened, and/or when electricity prices are very high. This approach generally limits customer interruptions to 8-20 times per year with duration limited to 2-6 hours each interruption. Participating customers can potentially rely on manual efforts (e.g., dimming lights, increasing thermostat set-points, shutting off equipment) to become a viable DR resource (although automation and control technology may increase the size and persistence of load curtailments from DR resources).

However, DR programs that provide ancillary services often require more frequent (e.g., several times daily) but much shorter (e.g., 10 minutes typically for spinning reserves) interruptions. Thus, in order to provide reserves and especially regulation, customers cannot rely on manual efforts to alter their consumption of electricity but rather must exclusively turn to control and automation technology to enable them to comply with control signals sent by the BA directly to the customer or via the program provider.

Most customers do not inherently possess this type of control technology; thus, the size and scope of the necessary technology investment depends upon several factors including: how many of the customers' electricity consuming devices are to be controlled, what types of devices are to be controlled, and how fast the response will need to be to provide the ancillary service. There may also be ongoing costs associated with various types of control and automation technology.

For example, the increased cycling of residential air conditioners or commercial chillers may increase the wear and tear on this equipment increasing operations and maintenance costs. These types of infrastructure investments may enable a customer to provide certain types of ancillary service, but additional costs must be incurred to adequately qualify as an AS provider.

Table 3. Rules that limit the magnitude of the DR resource

	Min. Size (MW)	Aggregation Allowed	Symmetric Regulation Capacity Req'd
ERCOT	0.1	No*	No
MISO	1.0	No	Yes
PJM	0.1	Yes†	Yes
PSCo	N/A	N/A	N/A

* Pilots are underway to examine the ability to change this rule.

† Requires approval before it can be implemented.

ISO/RTO requirements to be an AS provider extend to the infrastructure that a resource must have to record its electricity usage on a time-scale consistent with the service being offered, to demonstrate its compliance with a dispatch signal and to transmit this information back to the system operator via a communications network. ISO/RTOs require an ability to measure a resource's performance as a bulk power system service provider. For services that don't require real-time visibility of response (i.e., capacity, energy and in some cases supplemental reserves), an interval meter at a customer's premise whose data can be polled and submitted to the ISO/RTO in bulk on a delayed basis has been deemed acceptable. For example, NYISO requires program providers to submit hourly meter reads of participating customers within 75 days after a declared emergency/capacity event (NYISO 2012).

As the speed of response and system impacts associated with inaccurately measuring that response grows (i.e., spinning reserves, regulation), additional technology is typically required to accommodate more detailed, timely and transparent measurement and verification efforts under current rules. For example, resources that provide regulation have traditionally been required to possess technology which allows system operators the ability to follow their performance on a highly time-differentiated scale (i.e., 2-5 second telemetry). Some ISO/RTOs, like CAISO, require resources to transmit that information over a dedicated communications network that directly connects the DR resource or its program provider to the BA's system operations center. Customers who take service from a utility that has invested in Advanced Metering Infrastructure (AMI) already have digital interval meters installed but this equipment often times does not support real-time data transmission nor capture electricity consumption at 2-5 second intervals; thus additional equipment must likely be invested in to qualify as an AS provider. Telemetry equipment constitutes an additional large upfront investment with costs tending to increase with the speed of response required. Two of the three ISO/RTOs in our study have telemetry requirements for spinning reserves and all three ISO/RTOs require it for regulation.[7] Although not required in MISO, PJM or ERCOT, procuring a dedicated communications network results in both a capital cost to set up the network but also monthly service charges to maintain and use that network.

Taken jointly, these enabling infrastructure costs may serve as a sizable economic barrier to new entry of smaller customers, if it is required for each and every customer load that is being aggregated by a DR resource provider. These monitoring and verification requirements were developed when all AS were provided by large generators and they may not be technically justified for aggregations of small responsive loads. This particular barrier can be somewhat mitigated if the telemetry rules for DR resources are relaxed, as some ISO/RTOs have done, or the costs may be reduced if state entities subsidize the cost of this equipment to customers.

Table 4 indicates a sampling of market rules in the case study regions that impact part of the cost of enabling infrastructure. PJM has the most favorable rules for DR resources of the three ISO/RTO regions in our study which impose the least cost burden to provide various forms of AS. They allow aggregate data to be reported, their telemetry may be batch sent once each minute, and they do not require telemetry for spinning reserve.

7. Revenue Availability Barriers

An analysis of the revenues for the two most valuable ancillary services (i.e., regulation and spinning reserves) can be used, in conjunction with explicit enabling infrastructure costs, to establish the payback period that could be supported by program providers looking to break into such markets.[8]

We focus our efforts here on characterizing the level and variability of revenue that could have been derived in our three ISO/RTO study regions as well as the lone non- ISO/RTO region over the past several years, in order to present the boundary conditions for such financial calculations that a program provider would need to undertake when assessing the robustness of these markets to new entry.

Table 4. Market rules impacting the cost of participation for DR resources

	Telemetry Rate	Telemetry for Spinning Res	Data Source Level	Dedicated Network Requirement
ERCOT	3-5 sec	Yes*	Aggregate	No
MISO	4 sec	Yes	Resource	No
PJM	2 sec†	No	Aggregate	No
PSCo	N/A	N/A	N/A	N/A

* Required for some resource types.
† May be batch sent every one minute.

7.1. ISO/RTO Regions

Organized wholesale markets run by ISO/RTOs provide a transparent source of valuation for DR resources and AS program providers on a temporal basis. The ancillary services market clearing price (MCP) that ISO/RTOs produce can be used as a proxy for the revenues that new market entrants could expect to receive in the near term. When analyzed jointly with the volume of capacity procured in an ISO/RTO market, DR resources and program providers can assess how much potential revenue is at stake for specific AS products. However, the potential impact of entry by several new low cost resources (e.g., certain types of demand response) on AS market prices in ISO/RTO environments is uncertain and may have the effect of lowering the clearing price (Woychik 2008).

7.1.1. ISO/RTO Market Clearing Price

The market clearing price of an ancillary service product is derived by a balancing area's system optimization algorithms (i.e., market software) and is basically equal to the sum of an availability bid and opportunity cost (generally determined through co-optimization with the energy markets) of the marginal supplier providing ancillary service capacity for each scheduled time period. In most ISO/RTO markets, all awarded resources are paid the MCP. MCP is recorded in units of $/MW-h, where MW-h represents the value of a MW of regulation or spinning reserves capacity held in reserve for an hour.

In the regions included in our study, the annual average MCP for regulation (up and down in ERCOT) ranged between ~$11/MW-h and

$31/MW-h from 2009-2011(see Table 5). However, with the exception of ERCOT, prices have been steadily declining during this time period. Annual averages mask seasonal and weather effects (e.g., high prices in the spring due to large amounts of water run-off and absence of hydroelectric plants in the AS market) which may have caused this observed time trend (MacDonald et al. 2012).

Thus, it is difficult to predict if prices will continue to drop over the next several years or will rebound to their 2009 levels. Such knowledge is vitally important to a DR resource or AS program provider who is considering entering such a market. If we assume that such prices are representative of those that could be achieved in the future and a program provider can construct a portfolio of DR resources that can provide regulation services for all hours of a 30-day month, these MCP values could yield revenues of $7.80 to $22.50 per kW-month. This reflects a difference in revenues of over 150% between markets but as little as 20% within a specific market (i.e., MISO) on an annual basis. Such uncertainty could make program providers shy away from entering certain markets, especially if it dramatically increases the risk that a program provider won't be able to enter into contracts with program participants for a period that allows them to fully recover the enabling infrastructure investment costs while making a profit. This is a particular concern in the state of Texas, given the variability in ERCOT's market prices and the typical duration of retail energy service contracts that have been observed in the Texas retail market.[9]

Table 5. Annual average market clearing prices for regulation and spinning reserves in U.S. ISO/RTOs

ISO (Reserve Zone)	Regulation (Combined or Up/Dn) ($/MW-h)			10-min Spinning Reserves ($/MW-h)		
	2009	2010	2011	2009	2010	2011
ERCOT	9.70/ 7.25	9.81/8.27	22.67/8.58	9.95	9.09	22.92
MISO	12.43	12.17	10.83	4.03	4.02	4.03
PJM*	23.51	17.95	16.42	4.83	5.72	7.91

* MCP for spinning reserve is for Mid-Atlantic reserve zone in PJM, while regulation MCP is for all of PJM Darker cell shades indicate real-time markets, lighter are day-ahead markets.

The annual average MCP for spinning reserve among ISO/RTOs is between $4/MW-h and $10/MW-h. Although these prices have stayed the same or increased between 2009 and 2011, as with regulation prices, these annual average values hide what may be short-term trends. For example, in 2011, the ERCOT region experienced brief periods of high prices that produced an anomalously high annual average of nearly $23/MW-h (MacDonald et al. 2012). Assuming a portfolio of resource could provide this service every hour in a 30 day month, this range in the value of spinning reserve would produce a monthly revenue stream of $2.90 to $7.20 per kW-month.

7.1.2. ISO/RTO Market Capacity Volume

Capacity volumes in ISO/RTO administered ancillary services markets vary by the size of a balancing area and how the balancing area determines its reserve requirements. Some BAs determine their requirements based on the forecasted load, while others base it on time of day, to reflect changing needs. Table 6 lists the average hourly capacity procurement in markets examined in this study. This data represents only the capacity that was purchased in the markets, and does not include any capacity that was self-scheduled or contracted bilaterally by market participants, for which data is not publicly available. The capacity reported for PJM's spinning reserve market is represented by the Mid-Atlantic Reserve Zone, which includes the State of New Jersey, while all other volumes reflect a market-wide procurement. On a capacity basis, ERCOT holds the most spinning reserve procured in its markets and PJM has the largest amount of regulation. But on a relative basis, when normalized by the system's 2011 average load, ERCOT procures the greatest share of both regulation (1.6% for either regulation up or regulation down) and spinning reserves (4.5%) of the three ISO/RTOs included in our study.

The variation in the relative size of the market volume is likely due to several factors. The ability for loads and generators to self-schedule AS outside of the various markets may help explain why MISO and PJM have such small shares of their total capacity procured to provide regulation and spinning reserves. In addition, ERCOT's electric isolation from the larger two US interconnections makes it more vulnerable to rapid frequency loss during contingency events resulting in a requirement to carry larger amounts of ancillary services.

In spite of these relative differences, the ancillary services markets are still very small when compared to the wholesale energy markets in these ISO/RTOs.

Traditional economic logic would suggest that a small market is unlikely to support a large number of new market entrants without substantially reducing the price.

Table 6. Average hourly in-market capacity procurement volume for some U.S. ISO/RTOs from 2009-2011

	Regulation (Combined or Up/Dn)		10-min Spinning Reserves	
	MW-h	% of Ave. Demand	MW-h	% of Ave. Demand
ERCOT	628/606	1.6%	1715	4.5%
MISO	396	0.6%	978	1.5%
PJM	824	1.0%	418	0.5%

Darker cell shades indicate real-time markets, lighter are day-ahead markets.

However, due to the co-optimization of energy and ancillary services, it is possible that the ancillary services MCP will be fairly robust to additions of new resources seeking to provide these services. MCP for ancillary services is largely driven by the opportunity cost of lost energy revenues. As the ancillary service market is only approximately 1% of the energy market, AS market clearing prices have limited impact on the overall system optimization. Thus, the opportunity cost component of MCP will likely remain constant even with large amounts of zero cost AS resources in the system. In fact, evidence of this exists currently, as ERCOT has approximately 50% of its synchronous reserve market satisfied by load, yet still has some of the highest MCPs for spinning reserve in the country. However, there will likely be a tipping point at which low cost resource penetration will cause MCP to begin to fall precipitously. Where that point exists, however, is not known nor has research been done in an attempt to identify it.

7.1.3. ISO/RTO Market Size

The market size gives an indication of how much money is available for potential DR program providers looking to develop a business model around offering ancillary services. Using the data for MCP and the capacity procurement volumes, an estimate of annual market size in dollars was calculated for the sample ISO/RTOs. Table 7 displays the annual market size of regulation and spinning reserve markets in the US in millions of dollars per year.

The market size for 2009 for MISO could not be calculated because of incomplete data. The data suggests that the deepest AS market for spinning reserve is in ERCOT, even without the spike in 2011 which was largely due to a single month's extremely high prices, and for regulation in PJM, although ERCOT is relatively close behind.

7.2. Non-ISO/RTO BAs

Unfortunately, a comparable comprehensive analysis of revenue for being an ancillary service provider in non-ISO/RTO BAs is not possible. Although the prices for the different ancillary services are available in the BA's FERC-approved tariff, the quantity that the BA provides to cover that fraction of the system requirement that is neither self-supplied nor forward-contracted for is not publicly available. This is the most comparable market volume metric to what we report for the ISO/RTOs. These non ISO/RTO BAs do not draw from a pool of external resources to meet a transmission customer's shortfall, but instead supply it from its own set of internal resources. Thus, in these non-ISO/RTO market environments, it is more likely that a vertically integrated utility would come to the BA with a DR program of their own for which they want credit against their AS requirements.[10] Because this volume of AS is completely outside of our scope of knowledge, we limit our assessment in non-ISO/RTO BAs to the level of prices paid for different forms of ancillary services. BAs in non-ISO/RTO market areas annually submit a cost-of-service study to FERC to justify the rates they want to charge to transmission customers for use of the bulk power system as well as for procuring ancillary services on their behalf should they arrive without a balanced schedule. Although WECC doesn't allow DR resources to currently provide regulation, as previously mentioned, it is still illustrative to see what rates were filed by PSCo and approved by FERC for both regulation and spinning reserves in order to assess how much revenue could be available for a demand response resource wishing to provide these services. As Table 8 shows, the price in 2010 and 2011 for regulation (combined up and down) were identical ($16.202/MW-h) as were prices for spinning reserves ($16.526/MW-h).

Data for 2009 was not available. These prices likely serve as the maximum, in the short-run, that a utility-sponsored DR program would offer their participating customers to provide the various ancillary services.

Table 7. Annual market size of U.S. ISO/RTO regulation and spinning reserves markets

[M$/yr]		ERCOT	MISO	PJM
Regulation	2009	105	NA	160
	2010	118	43	126
	2011	152	38	123
Spinning Reserve	2009	119	NA	24
	2010	122	33	32
	2011	462	23	51

Darker cell shades indicate real-time markets, lighter are day-ahead markets.

Table 8. PSCo Annual hourly point-to-point delivery price

	Regulation (Combined		**10-min Spinning Reserves**	
Year	2010	2011	2010	2011
Price ($/MW-h)	16.202	16.202	16.526	16.526

8. REVENUE CAPTURE BARRIERS

While an analysis of these wholesale and bilateral markets illustrates that there is money to be earned from providing various forms of ancillary services, a variety of barriers may exist that limits the ability for entities offering AS programs or DR resources participating in them to sufficiently capture this revenue to make the opportunity worth pursuing (see Table 9).

The uncertainty surrounding the availability of DR resources to provide ancillary service during different points in time and the measurement of their performance when scheduled lead to uncertainty in how much revenue a resource can expect to capture for providing these services. Some strategies that customers employ as a DR resource are available every hour of every day to provide certain types of ancillary services, but most have restrictions at different times of the day, days of the week, etc. that limits their availability to be an AS supplier.

When coupled with the high volatility in market clearing prices that has been observed in some ISO/RTO ancillary services markets (see Table 5), predicting a reasonably accurate revenue stream over time that DR resources and their program providers in these jurisdictions might capture becomes challenging. This is exacerbated when considering that most methods used to measure demand response resources rely on meter data that includes both controllable loads, which are used to provide the bulk power system service, and uncontrollable loads, which are not and therefore subject to variability during the operating period. For some types of customers, divorcing the metering of controllable and uncontrollable loads at a site, thereby effectively sub-metering the loads that will respond to market control signals, should improve the accuracy of the measurement. This reduces revenue uncertainty but may increase enabling infrastructure investment costs.

Some of the ways in which DR resources interact with the ISO/RTO to provide a specific ancillary service can further inhibit their ability to capture certain kinds of revenue streams by limiting the quantity of a service the DR resource provides and/or is paid for. For example, a DR resource may be functionally capable of providing more than one bulk power system service (e.g., installed capacity, spinning reserves and regulation), but market rules may limit that DR resource to only provide a single service or restrict the size of the resource that can qualify to provide different services. Removing or relaxing these rules at the ISO/RTO may increase the amount of demand response that can be brought to market.

Additionally, some system operators require DR resources to jointly bid into energy and ancillary services markets even if that resource only has the intention of providing an ancillary service. If the resource cannot economically provide energy for certain dispatch periods if called upon, the DR resource must either hedge itself against the risk of being dispatched with some sort of financial instrument, take itself out of the market for all bulk power system services during such times, or submit a very high energy bid in an attempt to prevent the system operator from dispatching it for energy but taking it as an ancillary service provider. If this requirement did not exist, the DR resource could fully bid its capability into the AS market at all times. Furthermore, in some jurisdictions (e.g., ERCOT), DR resources are not paid directly for the energy they provide the system when acting as an ancillary service provider. An inability to capture this revenue stream, which generators are provided with, may make it more difficult for a DR resource or an AS program provider to justify investments in enabling infrastructure. Interestingly, the DR resource's supplier of electricity will actually capture much of this benefit, not the DR resource itself, if the wholesale price of electricity is substantially higher than the customer's retail rate. Lastly, certain kinds of demand response can provide a fast and flexible resource capable of providing AS with potentially insignificant startup costs or startup time. Aside from meeting the basic bulk power system service definitions, providing a higher quality service based on such attributes is not valued in most current market designs. However, FERC, with its Order 755 (FERC 2011b) changed that by requiring organized wholesale markets to adopt methodologies where speed and accuracy of response will directly affect revenue streams. This, in turn, will likely improve the financial viability for DR resources and AS program providers.

Since it will most likely be vertically integrated utilities in these non-ISO/RTO BA environments that come forward with DR programs that provide AS, many of these barriers are more easily circumvented. For example, as long as there is a financial or regulatory incentive to do so, a vertically integrated utility can decide to allow or require sub-metering if it feels such is justified to more accurately measure the ancillary service a DR resource is providing. Likewise the utility can decide how best to dispatch its energy resources in these types of environments, thereby not requiring DR resources providing AS to offer up any energy.

Additionally, the BA would not impose any obligations on having its DR resources be paid for any energy that is supplied; rather the utility's tariff would define what to do in such circumstances. Unless the utility can capture the value created when DR resources provide energy, the IOU would have no source of revenue to pay the participating customer for this energy, absent going out to its ratepayers to collect the money. The lone barrier in Table 9 that would likely directly apply is a multi-program restriction. The BA could require the utility to certify that its DR resources are not being credited against multiple bulk power system service requirements, but such rules have not been implemented by PSCo yet to our knowledge.

9. PROGRAM PROVIDER BARRIERS

The structure and design of retail electricity markets also play a large role in influencing participation and interest in demand response, given the incumbent utility's unique role as the dominant interface between the wholesale power grid and the end-use customer. However, under traditional regulation, electric utilities have a regulatory model that itself may create barriers that limit their interest in pursuing DR resources as an ancillary service provider. Aggregators of retail customers, whose business operations are predicated on subscribing electric customers to demand response programs, have barriers of their own that limit their interest and ability to seek out DR resources willing to provide ancillary services.

Table 9. Status of rules affecting revenue capture

	Allow Submetering	Must Bid Energy	Paid for Energy	Multi-Program Restrictions
ERCOT	No	No	No	No
MISO	No	Yes	Yes	No
PJM	Yes	No	Yes	No
PSCo	N/A	N/A	N/A	N/A

9.1. Retail Electric Utility Environment

The traditional regulatory model for monopoly investor-owned electric utilities substantially drives their decision making surrounding most areas, demand response included. State regulatory commissions set and approve retail rates for IOUs through a regulatory process (i.e., rate case) designed to allow them to recover their prudently incurred costs and generate enough profit to gain access to capital by offering a reasonable rate of return on existing capital investments. Since these retail rates, for the most part, do not change considerably between rate cases, the utility has an incentive to reduce operating expenses and/or increase revenue (by promoting greater volume of retail electricity sales) in order to generate additional profit for its shareholders. Undertaking large capital investments does carry some risk, as regulators must inevitably approve what is eligible for cost recovery during a rate case, but also carry a reward, as regulators allow a rate of return on all acceptable capital expenditures at the utility's authorized return on equity which will increase the utility's allowable profit that rates are set to recover. Given this regulatory paradigm, IOUs have a business model that is not readily conducive for the pursuit of activities, such as energy efficiency or demand response, that will increase costs, reduce revenues, or defer future capital investment (see NAPEE 2007 for more details).

The history of demand response in IOUs illustrates the challenges this business model produced but also how utilities overcame them. In the 1970s, large industrial customers, seeking to reduce their own costs, demanded retail electric rates that would promote economic development. Out of concerns about the impacts such a large loss of these customers' load would have on utility revenues, IOUs developed the earliest forms of demand response–interruptible and curtailable (I/C) rates. These were designed to improve reliability, by giving the utility an opportunity to infrequently reduce electric service at a customer's facility (e.g., 20 times a year at most), while substantially reducing that customer's retail rate for that amount of load what was "non-firm" or interruptible. Regulators saw the benefit, both to the utility and its system from the increased reliability but also to the industrial customers by way of economic development, and were willing to approve the rate and allow the utility to recover any prudently incurred costs to implement it. Later that century, public policymakers and regulators wanted utilities to broaden their demand side management efforts (i.e., energy efficiency and demand response) as a way to further improve reliability, reduce system costs, and contribute to environmental sustainability. As a strategic response to this request, utilities sought to again pursue demand response activities (i.e., direct load control or DLC programs) provided regulators would authorize full and timely recovery of program costs and enable them to generate a return on the investment in control technology that they installed via retail rates. Under these conditions, such DR programs did not hurt the utility's bottom line nearly as much as energy efficiency efforts would, as they allowed the utility to retain load (thereby not reducing revenue), ratebase the investment in DLC technology (thereby earning a profit), while providing the utility with the ability to credit the peak demand reductions against planning reserve requirements (thereby reducing overall system costs). During the period between 1970s and mid-1990s (prior to restructuring), many utilities across the US brought forth such programs because they had solved the fundamental problem with the existing business model or facilitated retention of key customer loads (e.g. large industrials). With advancements in technology over the past ~20 years coupled with increasing interest of customers to have some ability to override a control signal, utilities began to move away from traditional DLC devices towards more flexible and customer-controllable technologies, like programmable controllable/communicating thermostats, to provide these types of demand response opportunities. It has now become difficult for utilities to claim these devices should qualify as a utility capital investment, since customers could readily pursue this technology on their own at relatively low cost, and that the utility should continue to be the exclusive program provider, as technology (e.g., AMI) had created opportunities for outsourcing and the advent of wholesale electricity markets have created opportunities for third-party DR program providers to emerge as a viable competitor. As such, the traditional utility approach to such forms of DR is being substantially challenged at present.

Historically, interruptible/curtailable tariffs and DLC programs were activated during system emergencies and in some cases, to respond to high prices in energy markets. However, IOUs face a somewhat different set of challenges when assessing if a business case exists for creating and offering either expanded or new DR programs that target reserves and regulation services. DR programs that provide ancillary services will produce financial benefits, but such benefits are difficult for the utility to capture. For example, if the provision of operating reserves from customers allows a vertically integrated utility to more efficiently operate its fleet of generation assets, the utility's operating expenses can be reduced in the form of lower fuel and purchased power budgets as well as reduced non-fuel operations and maintenance costs. Any reduction in operations and maintenance costs would go to the utility's bottom line but only up until the next general rate case is filed and new rates reflective of these lower costs go into effect. Savings in fuel and purchased power budgets from DR resources providing AS, which would occur if the utility uses its own generation assets to provide ancillary services or must go out to the market to procure it, are generally passed directly through to customers via a fuel adjustment clause, leaving the utility with little to no ability to profit from any reductions in these costs. However, should these types of DR programs result in an opportunity to convert that reserved generation capacity into energy and sell off-system at a profit, regulators have struck various deals with utilities that require some, most, or all of the net proceeds to be returned to ratepayers. For example, Wisconsin does not allow the utility to retain any profit from an off-system sale (Wisconsin Administrative Code 2012), whereas Public Service of Colorado can retain up to 20% of the net proceeds (Colorado PUC 2009). Lastly, increased demand for ancillary services may necessitate the building of new generating stations in the future to accommodate this need. If DR resources are instead relied on to provide such services, the need to invest in this new capacity would be deferred or mitigated outright thereby causing the utility to forego a potential profit opportunity.

Aside from business model issues, the incumbent retail electric utility has other challenges offering these types of DR programs. As stated above, traditional forms of demand response, like those predominantly and historically offered in Colorado, Texas, Wisconsin and New Jersey, have focused on providing bulk power system services that exclusively inure from reducing electricity consumption (e.g., DLC, I/C) just a few times each year with duration of only a few hours each time. DR program designs that target ancillary services markets will need to fundamentally change to accommodate more frequent but shorter duration events and may require customers to both reduce and increase load, in the case of frequency regulation, as they follow dispatch signals from the ISO/RTO's or the BA's control center. Although the system operator in the control center is the originator of such control signals, a clear conflict of interest arises when a utility, who is managing and subscribing customers to such programs, could profit from the increased consumption of electricity. Additionally, utility tariffs are designed to treat customers as a group and not allow the utility to price electric service differentially for each and every customer (i.e., price discrimination). In order to provide the BA with the appropriate amount of ancillary services being called for, the IOU may have to differentially dispatch DR resources such that some customers are called upon more or less than others and get compensated accordingly. This differential treatment poses challenges to the traditional regulatory compact under which utilities' generally operate.

Thus, state regulators continue to play an important role in allowing and inducing investor- owned utilities, either vertically integrated or facing retail competition, to pursue such DR opportunities. Tariffs must be changed that allow the utility to offer programs to customers that substantially alter performance requirements and dispatch conditions from previous DR programs. Regulators must also address the shortcomings in the current business model to induce utilities to pursue these types of programs. First, regulators can guarantee timely cost recovery of prudently incurred costs to design, implement and maintain these types of AS programs.[11] Additionally, they can continue the tradition of allowing the utility to earn a rate of return on the necessary enabling infrastructure investments that a customer must undertake to provide AS or alternatively find other methods for making this endeavor profitable for the utility. Lastly, regulatory commissions can consider different approaches for altering a utility's tariff to allow for it to capture more of the profit that may be generated from changes in the operations of its generation fleet (if it has not been divested of them) and to differentially dispatch and/or provide different performance payments for customers participating in a DR ancillary services program. To date, none of the states in this study have undertaken a comprehensive assessment of what it would take to induce utility's to provide these types of opportunities to customers, but based on our interviews Texas appears ready to do so in the near future.

9.2. Aggregators of Retail Customers

The formation of ISO/RTO wholesale markets created an opportunity for non-utility entities to see and capture value in subscribing customers to provide various forms of bulk power system services. However, these aggregators of retail customers must contend with their own set of unique barriers which can limit or hinder their ability to offer such programs.

State regulators can be uncomfortable relinquishing control over DR programs to private sector entities that are largely unregulated. The level of oversight given by state governmental bodies of ARCs is often times far less than state regulators impose on electric utilities under their jurisdiction. ARCs generally must conform to program rules set by ISO/RTOs that themselves are regulated by FERC.

One state regulatory staffer indicated when interviewed that this can create a conflict, whether perceived or real, between state and federal jurisdictions. Often times, ARCs do not supply electricity to their program participants yet their programs affect the timing and amount of electricity consumed. Regulators and utilities alike have concerns about the implications this can have on the efficient and economical procurement of energy for these customers.

Taken in totality, a few state regulators (e.g., Wisconsin) have outright prohibited the operation of ARCs; while in other states, regulators have required the utility to outsource DR programs to ARCs (e.g., Colorado); while still in others they are freely allowed to subscribe customers to their own programs (e.g., Texas and New Jersey). Such decisions that limit the opportunities for ARCs to offer programs to customers, however, may result in an inability for DR to reach its full potential if the utility is unable to overcome its own set of barriers to offer such DR programs (as discussed above).

In non-ISO/RTO BA environments, it is the BA itself (or its associated IOU) that is supplying the bulk power system services to transmission customers who arrive without a balanced schedule. As such, ARCs have no real buyer of the bulk power system services they may be able to offer, except a utility. Thus, they must work with the utilities within the balancing authority's footprint or have regulators require the ARC to either manage a DR program on behalf of the IOU or have that program outsourced to them entirely.

If ARCs are able to participate directly in ISO/RTO markets, they must find the endeavor sufficiently profitable in order to pursue it. Given the volatility in ISO/RTO AS market revenues (see Sections 7 and 8), an ARC in such jurisdictions must be able to lock in a customer for a sufficient length of time to ensure it can pay off whatever initial investment is made in automation and control technology that is necessary to enable a customer to provide the contracted-for ancillary service. For example, in ERCOT's territory, there is considerable uncertainty in customer retention for retail energy providers. Although some customers sign contracts with a service provider for as much as 24 months, most customers do not. There is a very real risk that customers will change providers before reaching the end of the payback period on a DR enabling infrastructure investment, which makes the decision to enter such markets challenging.

9.3. Inducing Customer Participation

For utilities or ARCs to provide ancillary services to a BA, they must inevitably recruit customers into a program and convince them to accept, if not directly invest in, the necessary automation and control technology required to provide ancillary services. This requires education, marketing and technical assistance to ensure that the customer is fully aware and capable of providing the ancillary service for which they are being paid. ARCs and utilities alike must overcome customer hesitancy surrounding the installation of systems that relinquish control of some electricity consuming devices to a third-party. Customers must be convinced that the program provider is acting in their best interest when dispatching them for an event and adjusting their consumption of electricity via the control technology.

Just as there is competition for customers between utility-sponsored and ARC-sponsored programs, there can be competition for customers between energy efficiency and demand response programs. In some applications, energy efficiency efforts can actually undermine demand response capabilities. The more these programs compete with each other and are not integrated or coordinated in their program development and delivery efforts, as is the case in Colorado, Wisconsin and New Jersey but not Texas (Goldman et al. 2010), the more challenging it will be to achieve robust DR and EE programs.

Additionally, certain program providers specialize in providing specific services (e.g., Enbala Power Networks® focuses on providing regulation services and more recently load following) or recruiting specific loads (e.g., EnerNOC, Inc. generally subscribes commercial and industrial customers), but a customer site might have many different types of controllable loads that are good candidates to provide different types of bulk power system services.

Customers should have the option of being represented by more than one program provider and/or enrolling in more than one DR program if they have a variety of controllable loads that are amenable to do so.

Customers also need to have enough money on the table to induce their participation. As noted above, the level of payments given for providing an ancillary service is both volatile and may decrease with increased participation. The cost of the enabling infrastructure investments to meet bulk power system service definitions may require an extended payback period when relying on the payments from participation in an ancillary services DR program alone. As such, a customer (or its program provider) may need to consider pairing enrollment in an ancillary service program with another DR program, to the degree such is allowed, that can leverage the technology investment. In the end, customers must see sufficient value both in the short and long-term, given the lengthy payback period for the enabling control technology required to provide such ancillary services, in order to subscribe to such programs.

CONCLUSION

Increasing penetration of variable renewable resources in US electricity markets is likely to require operational changes and procurement of greater quantities of various bulk power system services. Although these services have been traditionally provided by generators, demand response has been identified as one of several alternative resources that could also serve this need. Conceptual studies and actual demonstrations have illustrated that certain forms of DR resources are well-suited to provide AS to the grid due in part to their fast response, distributed nature and the statistical reliability of large numbers of smaller resources. DR resources may also have the ability to be brought to market much more quickly than a comparably-sized generation resource that must go through the lengthy permitting, siting and regulatory approval process.

However, the interest in fostering demand response participation in ancillary services markets requires the opportunity to provide such services, a clear financial incentive to do so and the market and regulatory structures to capture this value. This study identifies the various barriers standing in the way of more widespread deployment of demand response resources as a provider of ancillary services.

Table 10. Status of issues affecting program providers

	Opportunities for ARCs	Viable Utility Business Model	Coordination of EE and DR Activities
Texas	Yes	No	No
Wisconsin	No	No	Yes
New Jersey	Yes	No	Yes
Colorado	Limited*	Limited†	Yes

* ARCs currently are only able to act as a third-party DR program provider.

† PSCo is allowed to retain 20% of the net revenues from off-system sales associated with excess electricity production.

Table 11. Applicable entities and organizations responsible for and affected by barriers

	Reliability Council	BA	IOU	ARC	Utility Regulator	End-use Customer
Bulk Power System ServiceDefinitions	*	*, •	•	•		
Attributes of Performance		*	•	•		•
Enabling Infrastructure Investments		*	•	•		•
Revenue Availability		*	•	•		
Revenue Capture		*	•	•		•
Program Providers		*	•	•	*	•

* - Entity/Organization responsible for creating the barrier.

• - Entity/Organization affected by the barrier.

As Table 11 summarizes, regional reliability councils must first allow DR resources to provide ancillary services by defining them in such a way that IOU or ARC DR programs are not precluded, either implicitly or explicitly, from doing so.[12] Once the opportunity is there, BAs promulgate rules that define the infrastructure and performance attributes of a DR resource wishing to provide such services that are brought to market directly or via a program provider. These rules may create barriers, either physical or financial, that limit the ability of a DR resource or program provider to provide these services. When taken in conjunction with expected revenue streams and an ability to capture this revenue as an AS provider, barriers that relate to the cost effectiveness of these resources are revealed which can limit DR from reaching its full potential. The regulatory compact between utility and state regulators, along with other statutes and decisions by state policymakers, may also create barriers that program providers must overcome in order to pursue DR resources more vigorously as an ancillary service provider.

One of the most effective ways to remove these barriers is to alter the requirements imposed on DR resources wishing to provide AS (see Table 12).[13] By acknowledging that demand response is fundamentally different than a generator, many ISO/RTOs are currently finding ways to alter the requirements to provide these services such that the quality of the service they are procuring is maintained but the pool of resources that can provide it is expanded. In contrast, most non- ISO/RTO BA environments do not appear to be as far along in their attempts to better integrate non-traditional resources as ancillary service providers. Where DR resources are able to provide AS, advancements in technology through research and development efforts and/or increased market adoption will likely cause the cost of automation and control technology and other forms of enabling infrastructure investments to continue to drop making participation as an AS provider more cost effective. Increases in benefits, through market rule changes (e.g., scarcity pricing, reserve demand curves) can likewise contribute to an increase in the cost effective procurement of AS from demand response. Finally, altering the process by which program providers do business (e.g., changes in the utility's business model) should help facilitate increased interest on their part in pursuing DR resources as a viable provider of AS.

The four regions focused on in this study illustrate these various barriers and ways in which the regions have or are attempting to address them. Colorado has a vertically integrated retail utility environment that operates within a non-ISO/RTO BA footprint, where the regional reliability organization (WECC) allows DR resources to provide spinning reserves but not regulation. The existing wholesale market environment results in little to no opportunity for ARCs to directly solicit customers and bring DR resources forward to provide AS unless there is someone on the other end to buy their services. The BA does not provide such a "marketplace", so the ARCs must develop relationships with the utilities in the BA's jurisdiction in order to play some sort of role. The investor-owned utilities, as the sole provider of electricity service to customers, have only modest profit motives to pursue non-generation resources, like DR. Colorado utilities and state regulators have not placed expansion of DR to provide ancillary services as a high priority given the excess capacity situation currently experienced by the state's utilities. If the capacity situation tightens and/or the distribution system requires attention due to increased penetration of distributed variable renewable resources, which may occur later this decade, state regulators as well as the utilities could be more inclined to consider such new DR opportunities.

Texas, with its open retail market and ISO/RTO wholesale market, appears to have conditions that are highly conducive for DR resources to readily provide ancillary services. ERCOT has indeed attracted a substantial amount of demand response to its markets, enrolling ~2,400MW of capacity to provide spinning reserves (Patterson 2011). The vast majority of this is provided by large industrial facilities utilizing under-frequency relays that were installed through utility- sponsored instantaneous interruptible tariffs years prior to the advent of the organized wholesale market (Zarnikau 2010). While ERCOT has been very successful at operating with sizable amounts of ancillary services (i.e., up to 50%) provided by large DR resources, relatively few new and substantially smaller DR resources have been brought to market. Conversations with program providers and regulatory staffers illustrate the challenges that the open retail market creates when trying to attract these smaller DR resources. The distribution utilities, who no longer have supply obligations to customers, have little profit motive to pursue such programs, which means regulators must get involved to either create the business case or mandate that programs be offered. Competitive retail electricity suppliers do not yet readily see a value proposition in offering the requisite types of enabling technology as an additional service due in part to short customer contract lengths which do not exceed the return on investment hurdle rates on the equipment. The current wholesale "energy-only" market design does not provide a reservation payment that guarantees a longer term (e.g., quarterly, semi-annual, or annual) revenue stream compared to other ISO/RTOs that administer capacity markets. This adds risk into the decision to invest in enabling infrastructure at customer premises. Reserve margins that exceeded ERCOT requirements by 2-4 percentage points coupled with major changes at ERCOT to a nodal market over the last several years resulted in time and effort not being directed towards creating greater opportunities for DR resources to participate in the market. Although with more recent reserve shortages, ERCOT and the Texas Public Utility Commission are jointly working through the stakeholder process to resolve these issues and improve the environment for DR in general (see Public Utility Commission of Texas Project No. 40000 for more details).

New Jersey is in a similar situation as Texas but for somewhat different reasons. Like ERCOT, PJM, the ISO/RTO that covers New Jersey, allows DR resources to provide spinning reserves and regulation. The existing utilities' business model, however, creates little incentive for distribution utilities to pursue new DR resources in any fashion, let alone as an ancillary service provider. New Jersey allows retail competition and utilities currently provide electric commodity and ancillary services as a bundled product for default service customers through a multi-year contracting (i.e., auction) process. The costs of this bundled product are completely passed through to these customers, which provides little incentive for the utility to pursue lower cost options to provide these services *outside of the auction process* nor any ability to do so within the auction process since the products are bundled. So even though PJM is on the forefront of creating and expanding opportunities for DR resources to provide various forms of ancillary services, it is only ARCs or competitive retail providers going directly to the market or through bilateral contracts with third-party suppliers[14] who are likely to bring resources to market, absent state regulatory engagement.[15]

Table 12. Actions required to overcome barriers

	Change Definition	Change Requirement	Change Process	Reduce Costs	Increase Benefits
Bulk Power System Service Definitions	♦				
Attributes of Performance		♦			
Enabling Infrastructure Investments		♦			
Revenue Availability					♦
Revenue Capture		♦			
Program Providers			♦		

♦ - Primary action to overcome barrier.
- Secondary action to overcome barrier.

Wisconsin, with its regulated retail environment and restriction on ARCs providing DR programs, faces both regulatory and utility business model barriers that limit their utilities' interest in pursuing DR resources as a provider of ancillary services within MISO. Electric utilities in this state are vertically integrated. Since they rely on their own generation assets to serve customers' needs, any reductions in non-fuel operating expenses from more efficiently operating this fleet of resources can be captured by the utility but only until new rates are set which reflect these now lower costs. Although these types of DR programs also create an opportunity for the state's utilities to convert reserved generation capacity into energy which can be sold off-system, state regulators do not allow the utility to retain any of these profits but instead require the utility to turn them over to ratepayers. Furthermore, the electric utilities profit from investing in capital assets, like new generating stations, which is often not true for DR investments. As such, Wisconsin utilities have only modest financial incentives to pursue DR in general, while state regulators have restricted ARCs from doing business in the state. Should state regulators choose to alter the utility's business model making the pursuit of DR resources as a provider of AS more lucrative, the utility must alter its existing tariff to expand the conditions under which current and future DR resources can be dispatched. Regulators could also choose to lift their moratorium on ARCs to further facilitate access to DR resources that can provide ancillary services. However, at this time, based on our interviews, the Commission does not appear inclined to look into these issues in the near future.

These four regions included as case studies in this report are likely representative of the experience in similar wholesale and retail market environments across the country. Our assessment of the barriers to demand response resources associated with smaller customers who must go through a program provider in order to provide ancillary services illustrates that the issues span nearly all entities and organizations in the chain that connects wholesale markets to retail customers.

Fundamentally, in a highly regulated system incentives must be aligned among the balancing authority, utilities, and program providers in order for DR to reach the system operator, where AS is procured. Most of these entities have little to no incentive to address the identified barriers on their own, even if society as a whole could greatly benefit from the effort. Thus, many different but disparate parties will need to work together for the common good in order for smaller DR resources to reach their full potential as a provider of ancillary services.

11. Appendix A. Entity/Organization Responsible for and Affected by Barriers

Category	Barrier	Consequence	Reliability Council	BA	Federal Utility Regulator	IOU	ARC	State Utility Regulator	End-Use Customer
Bulk power system service definitions	**Rules do not explicitly identify any form of demand response and/or storage resource as being eligible to provide any form of ancillary services**	Excludes all forms of demand response and/or storage resources from providing any form of ancillary service	*	*		•	•		
Bulk power system service definitions	**Rules explicitly identify specific forms of demand response (e.g., interruptible) and/or storage (e.g., pumped hydro) resources as being eligible to provide certain forms of ancillary services**	Excludes certain forms of demand response and/or storage resources from providing certain forms of ancillary services	*	*		•	•		
Attributes of Performance	**Rules impose certain size restrictions on magnitude of demand response and/or storage resources response capabilities (e.g., 1 MW, 0.1 MW)**	Excludes smaller forms of demand response and/or storage resources		*	*	•	•		
Attributes of Performance	**Rules impose certain geographic restrictions on participation (e.g., transmission zone) that are not technically warranted for reliability reasons**	Excludes aggregations of demand response and/or storage resources spanning different geographic locales		*	*	•	•		
Attributes of Performance	**Rules impose certain restrictions on timing of response capabilities (e.g., 6 second, 5 minute) that are not technically warranted for reliability reasons**	Excludes forms of demand response and/or storage response where communication latency and ramping requirement exceeds response time requirements (exacerbated if		*	*	•	•		

Category	Barrier	Consequence	Reliability Council	BA	Federal Utility Regulator	IOU	ARC	State Utility Regulator	End-Use Customer
		communications requirements include sending dispatch signals via TO, see communications barrier)							
Attributes of Performance	**Rules impose certain restrictions on duration of response capabilities (e.g., 30 minutes) that are not technically warranted for reliability reasons**	Excludes forms of demand response and/or storage response where duration of response requirement exceeds length of time the resource can provide service before being completely depleted		*	*	•	•		
Attributes of Performance	**Rules impose symmetric response requirements to provide a bulk power system service (e.g., identical capacity for regulation up and regulation down)**	Potentially limits the amount of the bulk power system service that could be provided by certain forms of demand response and/or storage		*	*	•	•		
Enabling Infrastructure Investment	**Rules impose certain restrictions on communications (e.g., security, network requirements, use of standard protocols, signal speeds for automatic generation control)**	Excludes forms of demand response and/or storage response where communication latency exceeds response time requirements		*	*	•	•		
Enabling Infrastructure Investment	**Complying with ancillary service event performance requirements may expose equipment to conditions that void warranties, reduces its useful lifetime, etc.**	Limits various demand response and/or storage resources from participating in ancillary services programs				*	*	*	•
Enabling Infrastructure Investment	**Rules impose certain restrictions on ability to directly receive and respond to dispatch signals (e.g., Automatic Generation Control) that are not technically warranted for reliability reasons**	Excludes aggregations of demand response and/or storage resources because of cost to comply with bulk power system product requirement aggregations of demand response and/or storage resources		*	*	•	•		

(Continued)

Category	Barrier	Consequence	Reliability Council	BA	Federal Utility Regulator	IOU	ARC	State Utility Regulator	End-Use Customer
Revenue Availability	**Benefits that can be captured are uncertain due to a variety of factors (e.g., high historic price volatility, thin A/S market volumes, level of penetration of renewable generation)**	Affects cost-effectiveness of effort to allow demand response and/or storage resources to provide ancillary services		*	*	•	•		•
Revenue Capture	**Rules impose restrictions on participants being subscribed by more than one program provider that are not technically warranted for reliability reasons**	Potentially limits the Balancing Authority from taking full advantage of all services a customer could provide if program providers don't offer programs that supply the full range of bulk power system services		*	*	•	•		
Revenue Capture	**Rules impose restrictions against participants providing certain combinations of bulk power system services (e.g., operating reserves and capacity) that are not technically warranted for reliability reasons**	Potentially limits the Balancing Authority from taking full advantage of all services a customer could provide if program providers don't offer programs that supply the full range of bulk power system services		*	*	•	•		
Revenue Capture	**Market rules impose restrictions against participants providing certain combinations of bulk power system services (e.g., operating reserves and capacity) that are not technically warranted for reliability reasons**	Limits interest in programs to only the most lucrative (i.e., cost effective) ones when a choice must be made between competing program options		*	*	•	•		•

Category	Barrier	Consequence	Reliability Council	BA	Federal Utility Regulator	IOU	ARC	State Utility Regulator	End-Use Customer
Revenue Capture	**Rules, in certain co-optimized systems, impose bidding requirements for both energy and various ancillary services**	Excludes forms of demand response and/or storage who are unable to cost-effectively provide energy for an extended length of time		*	*	•	•		
Revenue Capture	**Rules, in certain co-optimized systems, impose payment requirements that preclude reimbursement for energy and/or "mileage" provided by Ancillary Service resources**	Excludes forms of demand response and/or storage who are unable to cost-effectively provide such services without these payments		*	*	•	•		
Revenue Capture	**Accuracy of response to dispatch signal is not valued appropriately**	Limits demand response and/or storage resources from capturing total value provided to Balancing Authority, if fewer resources would have been required to provide bulk power system services		*	*	•	•		
Revenue Capture	**Baseline forecasting of demand has high uncertainty and a lack of accuracy**	Potentially affects performance calculations thereby affecting value capture for IOU/Aggregator who must subscribe large enough resources who can more accurately produce baselines or oversubscribe aggregations of resources to manage these inaccuracies		*	*	•	•		
Revenue Capture	**Rules impose certain restrictions on sub-metering that are not technically warranted for reliability reasons**	Potentially reduces accuracy in baseline estimates of usage which may affect performance calculations thereby affecting value capture for IOU/Aggregator		*	*	•	•		

(Continued)

Category	Barrier	Consequence	Reliability Council	BA	Federal Utility Regulator	IOU	ARC	State Utility Regulator	End-Use Customer
Program Provider	**Aggregators required to bid their demand response and storage resources participating in programs into wholesale markets via the participating customers' Load Serving Entity**	Limits Aggregators from offering such services		*	*	•	•		
Program Provider	**IOUs unable to differentially dispatch and pay customers on the same ancillary service program**	Precludes IOUs from offering any ancillary services programs to aggregations of demand response and/or storage resources				•		*	
Program Provider	**States exclude Aggregators from running their own programs, independent of the IOU**	Precludes Aggregator from offering any ancillary services programs to any form of demand response and/or storage resources					•	*	
Program Provider	**State regulators do not support a third-party DR program provider model**	Precludes Aggregator from offering any ancillary services programs to any form of demand response and/or storage resources		•			•	*	
Program Provider	**Software and back-office systems to dispatch and process demand response and/or storage resources as ancillary services providers must be authorized by applicable (regulatory) authority**	Affects interest in ISO/RTO/BA and IOU to pursue demand response and/or storage resources to provide ancillary services due to risk of cost-recovery		•	*	•		*	
Program Provider	**Ancillary services costs are generally a pass-through to ratepayers**	Limits interest in overcoming barriers to allow demand response and/or storage resources to provide ancillary services		•		*		*	•

Category	Barrier	Consequence	Reliability Council	BA	Federal Utility Regulator	IOU	ARC	State Utility Regulator	End-Use Customer
Program Provider	**Reliance on demand response and/or storage resources as ancillary service providers may reduce total sales**	Limits interest in overcoming barriers to allow demand response and/or storage resources to provide ancillary services		•		*		*	•
Program Provider	**Limited profit motive in short-run for pursuing demand response and/or storage resources as ancillary service providers**	Limits interest in overcoming barriers to allow demand response and/or storage resources to provide ancillary services		•		*		*	•
Program Provider	**Reliance on demand response and/or storage resources as ancillary service providers may reduce profits in long-run (i.e., defer need for adding new generation resources)**	Limits interest in overcoming barriers to allow demand response and/or storage resources to provide ancillary services		•		*		*	•
Program Provider	**Frequency and duration of ancillary services events that are not technically warranted for reliability reasons may exceed some customers willingness to participate**	Limits various demand response and/or storage resources from participating in ancillary services programs				*	*	*	•
Program Provider	**Complying with ancillary service event performance requirements may cause electricity bill to rise due to retail rate design (e.g., ratcheted demand charge, real-time pricing)**	Limits various demand response and/or storage resources from participating in ancillary services programs				*	*	*	•
Program Provider	**Retail Energy Provider contracts in deregulated markets are too short to provide enough revenue potential to merit capital investments in DR enablement**	Limits interest in demand response, especially in enabling loads to provide more M+V intensive products				*	•	*	•

(Continued)

Category	Barrier	Consequence	Reliability Council	BA	Federal Utility Regulator	IOU	ARC	State Utility Regulator	End-Use Customer
Program Provider	**Rules impose certain financial/credit requirements (e.g., posting a bond)**	Excludes forms of demand response and/ or storage who do not have the financial resources to meet met the requirements		*	*	•	•		
Program Provider	**Program administration, program compliance (e.g,. telemetry, metering) and control technology (e.g., EMCS) costs to enable specific kinds of demand response and/or storage resources to comply with specific ancillary services reliability and market rules that are not technically warranted for reliability reasons is high**	Affects cost-effectiveness of effort to allow demand response and/or storage resources to provide ancillary services		*	*	•	•		•
Program Provider	**Benefits provided by DR and storage resources to the grid that can be captured, because externalities to the market are not reflected, are too low relative to costs to enable specific kinds of demand response and/or storage resources to comply with specific ancillary services reliability market rules**	Affects cost-effectiveness of effort to allow demand response and/or storage resources to provide ancillary services		*	*	•	•		•
Procedural	**Market rule changes to correct observed problems with market products at Balancing Authority level must successfully navigate stakeholder working group and committee structure**	Extends the time required to address identified barriers that preclude demand response and storage from effectively providing various forms of ancillary services		*	*	•	•		

Category	Barrier	Consequence	Reliability Council	BA	Federal Utility Regulator	IOU	ARC	State Utility Regulator	End-Use Customer
		Presents a free rider problem where the entity that does all the work to get rules changed provides benefits to all other potential DR suppliers							
Procedural	**Regulatory change to correct observed problems with state-level DR program offerings and/or business models for such programs must successfully navigate the regulatory process**	Extends the time required to address identified barriers that preclude demand response and storage from effectively providing various forms of ancillary services Presents a free rider problem where the entity that does all the work to get rules changed provides benefits to all other potential DR suppliers				*	•	*	
Procedural	**Market rule changes to correct observed problems with market products at Balancing Authority level must successfully navigate stakeholder working group and committee structure**	Potentially extends the time and cost required to address identified barriers that preclude demand response and storage from effectively providing various forms of ancillary services Presents a free rider problem where the entity that does all the work to get rules changed provides benefits to all other potential DR suppliers		*	*	•	•		•

Legend

* = Entity/Organization is responsible for creating the barrier;

• = Entity/Organization is directly affected by the barrier.

12. Appendix B – Action Required to Overcome Barriers

Category	Barrier	Consequence	Change Definition	Change Requirement	Change Process	Reduce Costs	Increase Benefits
Bulk power system service definitions	**Rules do not explicitly identify any form of demand response and/or storage resource as being eligible to provide any form of ancillary services**	Excludes all forms of demand response and/or storage resources from providing any form of ancillary service	■				
Bulk power system service definitions	**Rules explicitly identify specific forms of demand response (e.g., interruptible) and/or storage (e.g., pumped hydro) resources as being eligible to provide certain forms of ancillary services**	Excludes certain forms of demand response and/or storage resources from providing certain forms of ancillary services	■				
Attributes of Performance	**Rules impose certain size restrictions on magnitude of demand response and/or storage resources response capabilities (e.g., 1 MW, 0.1 MW)**	Excludes smaller forms of demand response and/or storage resources		■			
Attributes of Performance	**Rules impose certain geographic restrictions on participation (e.g., transmission zone)**	Excludes aggregations of demand response and/or storage resources spanning different geographic locales		■			
Attributes of Performance	**Rules impose certain restrictions on timing of response capabilities (e.g., 6 second, 5 minute) that are not technically warranted for reliability reasons**	Excludes forms of demand response and/or storage response where communication latency and ramping requirement exceeds response time requirements (exacerbated if communications requirements include sending dispatch signals via TO, see communications barrier)		■			
Attributes of Performance	**Rules impose certain restrictions on duration of response capabilities (e.g., 30 minutes) that are not technically warranted for reliability reasons**	Excludes forms of demand response and/or storage response where duration of response requirement exceeds length of time the resource can provide service before being completely depleted		■			

Category	Barrier	Consequence	Change Definition	Change Requirement	Change Process	Reduce Costs	Increase Benefits
Attributes of Performance	**Rules impose symmetric response requirements to provide a bulk power system service (e.g., identical capacity for regulation up and regulation down)**	Potentially limits the amount of the bulk power system service that could be provided by certain forms of demand response and/or storage		■			
Enabling Infrastructure Investment	**Rules impose certain restrictions on communications (e.g., security, network requirements, use of standard protocols, signal speeds for automatic generation control) that are not technically warranted for reliability reasons**	Excludes forms of demand response and/or storage response where communication latency exceeds response time requirements		■			
Enabling Infrastructure Investment	**Complying with ancillary service event performance requirements may expose equipment to conditions that void warranties, reduces its useful lifetime, etc.**	Limits various demand response and/or storage resources from participating in ancillary services programs			■		
Enabling Infrastructure Investment	**Rules impose certain restrictions on ability to directly receive and respond to dispatch signals (e.g., Automatic Generation Control) that are not technically warranted for reliability reasons**	Excludes aggregations of demand response and/or storage resources because of cost to comply with bulk power system product requirement		■		□	
Revenue Availability	**Benefits that can be captured are uncertain due to a variety of factors (e.g., high historic price volatility, thin A/S market volumes, level of penetration of renewable generation)**	Affects cost-effectiveness of effort to allow demand response and/or storage resources to provide ancillary services				□	■
Revenue Capture	**Rules impose restrictions on participants being subscribed by more than one program provider that are not technically warranted for reliability reasons**	Potentially limits the Balancing Authority from taking full advantage of all services a customer could provide if program providers don't offer programs that supply the full range of bulk power system services		■			
Revenue Capture	**Rules impose restrictions against participants providing certain combinations of bulk power**	Potentially limits the Balancing Authority from taking full advantage of all services a customer		■			

(Continued)

Category	Barrier	Consequence	Change Definition	Change Requirement	Change Process	Reduce Costs	Increase Benefits
	system services (e.g., operating reserves and capacity) that are not technically warranted for reliability reasons	could provide if program providers don't offer programs that supply the full range of bulk power system services					
Revenue Capture	**Market rules impose restrictions against participants providing certain combinations of bulk power system services (e.g., operating reserves and capacity) that are not technically warranted for reliability reasons**	Limits interest in programs to only the most lucrative (i.e., cost effective) ones when a choice must be made between competing program options			■		
Revenue Capture	**Rules, in certain co-optimized systems, impose bidding requirements for both energy and various ancillary services**	Excludes forms of demand response and/or storage who are unable to cost-effectively provide energy for an extended length of time		■			
Revenue Capture	**Rules, in certain co-optimized systems, impose payment requirements that preclude reimbursement for energy and/or "mileage" provided by Ancillary Service resources**	Excludes forms of demand response and/or storage who are unable to cost-effectively provide such services without these payments		■			□
Revenue Capture	**Accuracy of response to dispatch signal is not valued appropriately**	Limits demand response and/or storage resources from capturing total value provided to Balancing Authority, if fewer resources would have been required to provide bulk power system services		■			□
Revenue Capture	**Baseline forecasting of demand has high uncertainty and a lack of accuracy**	Potentially affects performance calculations thereby affecting value capture for IOU/Aggregator who must subscribe large enough resources who can more accurately produce baselines or oversubscribe aggregations of resources to manage these inaccuracies		■			

Category	Barrier	Consequence	Change Definition	Change Requirement	Change Process	Reduce Costs	Increase Benefits
Revenue Capture	**Rules impose certain restrictions on sub-metering that are not technically warranted for reliability reasons**	Potentially reduces accuracy in baseline estimates of usage which may affect performance calculations thereby affecting value capture for IOU/Aggregator		■		□	
Program Provider	**Aggregators required to bid their demand response and storage resources participating in programs into wholesale markets via the participating customers' Load Serving Entity**	Limits Aggregators from offering such services		■			
Program Provider	**IOUs unable to differentially dispatch and pay customers on the same ancillary service program**	Precludes IOUs from offering any ancillary services programs to aggregations of demand response and/or storage resources		■			
Program Provider	**States exclude Aggregators from running their own programs, independent of the IOU**	Precludes Aggregator from offering any ancillary services programs to any form of demand response and/or storage resources		■			
Program Provider	**State regulators do not support a third-party DR program provider model**	Precludes Aggregator from offering any ancillary services programs to any form of demand response and/or storage resources		■			
Program Provider	**Software and back-office systems to dispatch and process demand response and/or storage resources as ancillary services providers must be authorized by applicable (regulatory) authority**	Affects interest in ISO/RTO/BA and IOU to pursue demand response and/or storage resources to provide ancillary services due to risk of cost-recovery			■		
Program Provider	**Ancillary services costs are generally a pass-through to ratepayers**	Limits interest in overcoming barriers to allow demand response and/or storage resources to provide ancillary services			■		
Program Provider	**Reliance on demand response and/or storage resources as ancillary service providers may reduce total sales**	Limits interest in overcoming barriers to allow demand response and/or storage resources to provide ancillary services			■		

(Continued)

Category	Barrier	Consequence	Change Definition	Change Requirement	Change Process	Reduce Costs	Increase Benefits
Program Provider	**Limited profit motive in short-run for pursuing demand response and/or storage resources as ancillary service providers**	Limits interest in overcoming barriers to allow demand response and/or storage resources to provide ancillary services			■		
Program Provider	**Reliance on demand response and/or storage resources as ancillary service providers may reduce profits in long-run (i.e., defer need for adding new generation resources)**	Limits interest in overcoming barriers to allow demand response and/or storage resources to provide ancillary services			■		
Program Provider	**Frequency and duration of ancillary services events may exceed some customers willingness to participate**	Limits various demand response and/or storage resources from participating in ancillary services programs			■		
Program Provider	**Complying with ancillary service event performance requirements may cause electricity bill to rise due to retail rate design (e.g., ratcheted demand charge, real-time pricing)**	Limits various demand response and/or storage resources from participating in ancillary services programs			■		
Program Provider	**Retail Energy Provider contracts in deregulated markets are too short to provide enough revenue potential to merit capital investments in DR enablement**	Limits interest in demand response, especially in enabling loads to provide more M+V intensive products			□	■	
Program Provider	**Rules impose certain financial/credit requirements (e.g., posting a bond)**	Excludes forms of demand response and/or storage who do not have the financial resources to meet met the requirements		■		□	
Program Provider	**Program administration, program compliance (e.g,. telemetry, metering) and control technology (e.g., EMCS) that are not technically warranted for reliability reasons costs to enable specific**	Affects cost-effectiveness of effort to allow demand response and/or storage resources to provide ancillary services		■		■	

Category	Barrier	Consequence	Change Definition	Change Requirement	Change Process	Reduce Costs	Increase Benefits
	kinds of demand response and/or storage resources to comply with specific ancillary services reliability and market rules is high						
Program Provider	**Benefits provided by DR and storage resources to the grid that can be captured, because externalities to the market are not reflected, are too low relative to costs to enable specific kinds of demand response and/or storage resources to comply with specific ancillary services reliability market rules**	Affects cost-effectiveness of effort to allow demand response and/or storage resources to provide ancillary services					■
Procedural	**Market rule changes to correct observed problems with market products at Balancing Authority level must successfully navigate stakeholder working group and committee structure**	Extends the time required to address identified barriers that preclude demand response and storage from effectively providing various forms of ancillary services			■		
Procedural	**Regulatory change to correct observed problems with state-level DR program offerings and/or business models for such programs must successfully navigate the regulatory process**	Extends the time required to address identified barriers that preclude demand response and storage from effectively providing various forms of ancillary services			■		
Procedural	**Market rule changes to correct observed problems with market products at Balancing Authority level must successfully navigate stakeholder working group and committee structure**	Potentially extends the time and cost required to address identified barriers that preclude demand response and storage from effectively providing various forms of ancillary services			■		

Legend

■ = Primary action to overcome the barrier.

□ = Secondary action to overcome the barrier.

REFERENCES

Callaway, D. S. (2009). Tapping the energy storage potential in electric loads to deliver load following and regulation, with application to wind energy. *Energy Conversion and Management.*, *50*(5), 1389-1400.

Cappers, P., Mills, A., Goldman, C., Wiser, R. & Eto, J. H. (2012). An assessment of the role mass market demand response could play in contributing to the management of variable generation integration issues. *Energy Policy*, (48), 420-429.

Colorado, P. U. C. (2009). Decision No. C09-1446: Order Addressing Phase I and ECA Issues [Docket No. 09AL-299E].

ERCOT. (2012a). Section 4: Day Ahead Operations. Section in ERCOT Nodal Protocols. ERCOT (2012b) Section 6: Adjustment Period and Real-Time Operations. Section in ERCOT Nodal Protocols.

ERCOT. (2012c). Section 9: Settlement and Billing. Section in ERCOT Nodal Protocols.

Eto, J. H., Nelson-Hoffman, J., Parker, E., Bernier, C., Young, P., Sheehan, D., Kueck, J. & Kirby, B. (2012). The Demand Response Spinning Reserve Demonstration--Measuring the Speed and Magnitude of Aggregated Demand Response. Presented at System Science (HICSS), 2012 45th Hawaii International Conference on.

FERC. (2008). Order 719: Wholesale Competition in Regions with Organized Electric Markets. FERC (2009) A National Assessment of Demand Response Potential: Staff Report. Washington, D.C. June 2009.

FERC. (2010). Delegated Letter Order: Compliance Filing for Public Service Company of Colorado, Transmission Tariffs [Docket ER10-02070-000].

FERC. (2011a). Order 745: Demand Response Compensation in Organized Wholesale Energy Markets.

FERC. (2011b). Order 755: Frequency Regulation Compensation in Organized Wholesale Power Markets.

FERC. (2012). *Electric Power Markets: National Overview*. 2012.

GE Energy. (2008). *Analysis of Wind Generation Impact on ERCOT Anclillary Services Requirements: Final Report*. Prepared for Electric Reliability Councile of Texas. March, 2008.

Goldman, C., Reid, M., Levy, R. & Silverstein, A. (2010). *Coordination of Energy Efficiency and Demand Response*. January 2010. LBNL-3044E.

Helman, U. (2010). *Resource and transmission planning to achieve a 33% RPS in California–ISO modeling tools and planning framework*. Presented at FERC Technical Conference on Planning Models and Software.

IRC. (2012). *North American Wholesale Electricity Demand Response Program Comparison*, 2011 Edition.

ISO-NE. (2012). *Market Rule 1 - Appendix J: Alternative Technology Regulation Pilot Program*. Kiliccote, S., Piette, M. A., Ghatikar, G., Koch, E., Hennage, D., Hernandez, J., Chiu, A.,

Sezgen, O. & Goodin, J. (2009). *Open Automated Demand Response Communications in Demand Response for Wholesale Ancillary Services*. November 2009.

Kirby, B. (2006). *Demand Response for Power System Reliability: FAQ*. Oak Ridge, TN. December 2006. ORNL/TM 2006/565.

Kirby, B. (2007). *Load Response Fundamentally Matches Power System Reliability Requirements*. Presented at Power Engineering Society General Meeting.

Kirby, B. & Kueck, J. (2003). *Spinning Reserve from Pump Load: A Report to the California Department of Water Resources*. Oak Ridge National Laboratory, Oak Ridge, TN, ORNL/TM-2003/99.

MacDonald, J., Cappers, P., Callaway, D. & Kiliccote, S. (2012). Demand Response Providing Ancillary Services. Presented at Grid-Interop.

Makarov, Y. V., Loutan, C., Jian, M. & de Mello, P. (2009). Operational Impacts of Wind Generation on California Power Systems. *Power Systems, IEEE Transactions on., 24*(2), 1039-1050.

MISO. (2012). *Business Practices Manual: Energy and Operating Reserve Markets*. NAPEE (2007) Aligning Utility Incentives with Investment in Energy Efficiency.

NERC. (2009). *Accommodating High Levels of Variable Generation: Special Report*. Princeton, NJ. April, 2009.

NERC. (2010). *Distubance Control Performance*. BAL-002-1.

NJBPU. (2008). Order - In the Matter of Demand Response Programs for Period Beginning June 1, 2009 -- Electric Distribution Company Programs [Docket No. EO08050326].

NJBPU. (2010). Order - In the Matter of Demand Response Programs for Period Beginning June 1, 2009 -- Electric Distribution Company Programs (Proposed Extension of Programs) [Docket No. EO08050326].

NYISO. (2012). NYISO Installed Capacity Manual. Rensselaer, NY. January 2012. Patterson, M. (2011) Demand Response in the ERCOT Markets. Presented at DOE Demand Response Workshop, Palatine, IL. October 25, 2011.

PJM. (2012a). *PJM Manual 11: Energy & Ancillary Services Market Operations*. PJM (2012b) PJM Manual 12, Balancing Operations.

Todd, D., Caufield, M., Helms, B., Generating, A. P., Starke, I. M., Kirby, B. & Kueck, J. (2008). Providing reliability services through demand response: A preliminary evaluation of the demand response capabilities of alcoa inc. *ORNL/TM*. 233.

WECC (2007). *Operating Reserves*. BAL-STD-002-0.

Wellinghoff, J. (2009). Remarks by FERC Chairman Jon Wellinghoff at CAISO Stakeholder Symposium. October 7, 2009.

Wisconsin Administrative Code. (2012). Wisconsin Administrative Code. PSC §117.05. Wiser, R., Barbose, G. and Holt, E. (2010) Supporting Solar Power in Renewable Portfolio Standards: Experience from the United States. Berkeley, CA. October, 2010. LBNL- 3984E.

Woychik, E. (2008). Optimizing demand response: A comprehensive DR business case quantifies a full range of concurrent benefits. *Public Utilities Fortnightly.*, *146*(5), 52-56.

Zarnikau, J. W. (2010). Demand participation in the restructured Electric Reliability Council of Texas market. *Energy.*, *35*(4), 1536-1543.

End Notes

[1] In addition to ancillary services, many renewable integration studies (e.g., GE Energy 2008; Makarov et al. 2009; NERC 2009) have identified a need for greater ramping capability to track large but relatively slow changes in electricity production from variable renewable generation resource. Since this product has not yet been defined by NERC or any of the balancing authorities, we chose to focus on the bulk power system services that are currently defined.

[2] FERC's Order 719 (FERC 2008) required ISO/RTOs to eliminate several identified barriers to demand response by amending their market rules as follows: accept bids from DR resources in certain AS markets; allow Aggregators of Retail Customers to fully participate in the market unless prohibited by the relevant regulatory authority; and modify pricing mechanisms during system emergencies and/or operating reserve shortages. FERC's Order 745 (FERC 2011a) sought to further eliminate barriers to demand response by requiring ISO/RTOs to amend their market rules to fully compensate DR at the locational marginal-price for providing energy when certain conditions apply. FERC's Order 755 (FERC 2011b) adopted a two-part compensation method for all resources that provide regulation service by requiring ISO/RTOs to pay all resources that clear the regulation market a: (1) uniform capacity payment; and (2) a performance payment based on the accuracy of response to system control signals.

[3] Large customers (e.g., aluminum smelting) in most ISO/RTO environments can and often times do currently participate directly in the market as an ancillary service provider. As such, the barriers they faced bringing their capabilities to the bulk power system are somewhat different than those of smaller customers (e.g., retail office buildings) who must go through a program provider (i.e., IOU, ARC). For example, large customers usually have the requisite interval metering already installed, can provide load reductions that meet minimum size requirements, and can more readily afford to invest in the necessary telemetry requirements. As such, we are focusing in this study on the barriers standing in

the way of BAs gaining access to smaller DR resources, as this is the group of customers that is still largely untapped. Many of the barriers listed here do still apply to larger electricity customers willing and able to go directly to the BA, but specifying where and explaining why this is or is not the case is beyond the scope of this paper.

[4] Where wholesale market DR programs exist, utilities and ARCs (where allowed) provide opportunities for customers to subscribe to these ISO/RTO DR programs. For simplicity, we characterize this as the ARC or utility offering DR programs as a "program provider" even though they technically only offer a service to facilitate entry into the ISO/RTOs DR program.

[5] This barrier is not unique to DR but rather one faced by any new resource seeking to enter a market.

[6] The term "aggregation of resources" is used here in a different context than the term "aggregator of retail customers" (ARC). As previously defined, ARCs are non-utility entities that offer opportunities for customers to act as a DR resource that provides a bulk power system service. An "aggregation of resources" simply refers to the ability to pool a group of disparate customers into a single portfolio for purposes of providing a bulk power system service in a DR program. Generally, state regulators and other policymakers determine if "aggregators of retail customers (ARCs) are allowed to offer DR programs. ISO/RTOs dictate if an "aggregation of resources" is allowed and if so what the conditions are for creating such a portfolio of participating customers.

[7] Telemetry, in this context, refers to additional metering requirements that the ISO/RTO imposes in order to have rapid meter readings be sent directly to the system operator for resource visibility purposes in real-time.

[8] For reference, non-spinning reserves tend to be about 7% to 50% of the value of spinning reserves, depending on the ISO/RTO, while 30-minute supplemental reserves are less valuable than non-spinning reserves.

[9] Interviews with both program providers and Commission staff indicated that retail power contracts can range from one to 24 months, with most customers preferring shorter contract lengths.

[10] For example, PacifiCorp has been using its CoolKeeper (DLC) program to satisfy part of its WECC requirements for non-spinning reserves for several years (Woychik 2008).

[11] Regulators may question the value of spending ratepayer dollars to invest in the back office systems and other forms of technology required to operate such programs when another entity (i.e., ARC) who has already done so could offer such programs at a fraction of the cost. Outsourcing of these types of programs, therefore, could be a concern for IOUs.

[12] For a more comprehensive listing of specific barriers and the entities and organizations responsible for and affected by these barriers, see Appendix A – Entity/Organization Responsible for and Affected by Barriers.

[13] For a more comprehensive listing of specific barriers and the actions that can be overtaken, see Appendix B – Action Required to Overcome Barriers.

[14] Third-party electricity suppliers provide capacity, energy, and ancillary services as a portfolio to competitive electricity suppliers or the incumbent utility for default service. The goal of third-party suppliers is to provide the product at least-cost, which may entail pursuing DR as a resource within the portfolio.

[15] The New Jersey Board of Public Utilities (NJBPU) in 2008 did get involved by ordering the utilities to augment DR program payments for new resources participating in PJM's 2009 Interruptible Load for Reliability (ILR) program through a series of one-year pilots with a goal of fostering cooperation with ARCs (NJBPU 2008). Two years later, the NJBPU stopped all utility pilot programs due to a lack of funding; leaving ARCs to continue enrolling customers directly in PJM programs without any augmented payment (NJBPU 2010).

INDEX

A

access, 10, 13, 15, 29, 35, 42, 50, 65, 79, 85, 89, 96, 97, 110, 121, 139
accounting, 66
AD, 70
adjustment, 112
administrators, 62, 73
ADR, 32
advancements, 81, 111, 118
agencies, 52, 61
aggregation, 46, 69, 98, 139
air emissions, 4
alters, 96
ARC(s), 68, 76, 78, 79, 81, 82, 84, 87, 93, 98, 114, 115, 117, 118, 120, 122, 123, 124, 125, 126, 127, 128, 129, 138, 139
assessment, 52, 58, 78, 85, 88, 106, 113, 121, 136
assets, 5, 9, 17, 28, 41, 44, 45, 57, 60, 61, 75, 84, 91, 112, 120
authority(s), 11, 68, 77, 78, 79, 85, 87, 89, 93, 96, 114, 121, 126, 133, 138
automate, 33, 38, 39
automation, 2, 8, 27, 29, 30, 31, 33, 37, 80, 81, 93, 99, 115, 118
average costs, 21
aversion, 40
avoidance, 4

B

barriers, 10, 25, 51, 68, 78, 79, 80, 81, 82, 84, 85, 87, 88, 89, 91, 92, 93, 94, 95, 96, 97, 98, 107, 109, 114, 116, 117, 118, 120, 121, 126, 127, 128, 129, 133, 134, 135, 138, 139
barriers to entry, 80, 95
base, 1, 9, 53, 57, 104
benefits, 3, 4, 6, 23, 29, 34, 35, 42, 43, 51, 65, 68, 76, 81, 112, 118, 129, 138
bias, 18
Bilateral, 16
business model, 9, 81, 83, 84, 87, 91, 93, 105, 110, 112, 113, 118, 120, 129, 135
businesses, 25
buyer(s), 90, 91, 114

C

candidates, 41, 64, 115
CAP, 1, 18
capital expenditure, 110
capital intensive, 18
carbon, 18, 78, 87
case study(s), 65, 71, 80, 85, 88, 94, 96, 98, 101, 121
CEC, vii, 1, 3, 46
certification, 35
challenges, 16, 24, 25, 39, 83, 110, 112, 119

China, 16, 29, 42, 43
classes, 31, 80, 93
clients, 33
coal, 15
cogeneration, 42
commercial, 1, 7, 19, 22, 24, 27, 29, 35, 37, 38, 39, 43, 44, 53, 69, 99, 115
commodity, 83, 120
communication, 31, 32, 36, 37, 38, 42, 68, 122, 123, 130, 131
communication abilities, 42
communication technologies, 32
comparative analysis, 71
compatibility, 35
compensation, 69, 73, 138
competition, 10, 35, 65, 83, 91, 92, 94, 113, 115, 120
competitive process, 11
compliance, 18, 32, 44, 73, 100, 128, 134
computer, 98
condensation, 31
conditioning, 2, 9, 39, 40, 50
conflict, 113, 114
conflict of interest, 113
congress, 45, 52, 70
consolidation, 41
construction, 4, 29
consumer surplus, 18
consumers, 11, 35, 51, 68
consumption, vii, 3, 6, 14, 24, 27, 30, 31, 36, 37, 43, 50, 73, 97, 98, 99, 100, 112, 115
consumption patterns, vii, 3, 50
contingency, 104
conversations, 92
cooling, 40, 41
cooperation, 84, 139
coordination, 8
cost effectiveness, 118
covering, 96
CPP, 1, 22, 27, 50
customer preferences, 74
cycles, 22
cycling, 31, 97, 99

D

data center, 37, 41
demand curve, 81, 118
demonstrations, 116
Department of Energy, vii, 1, 2, 3, 13, 21, 45, 46, 70, 75
deployments, 30, 31, 37, 69
deregulation, 10
differential treatment, 113
directors, 91
discrimination, 113
distribution, 2, 4, 5, 16, 18, 19, 20, 28, 63, 74, 75, 82, 83, 91, 119, 120
District of Columbia, 11
diversity, 92
drawing, 51
DSM, 6, 7, 8, 16, 17, 28, 45

E

economic development, 110
education, 8, 45, 74, 115
effluent, 40
EIS, 61
emergency, 2, 10, 11, 12, 14, 15, 29, 34, 41, 54, 59, 62, 64, 66, 73, 100
emission, 17
employment, 14
end-users, 6, 29
energy consumption, 6, 22, 28
Energy Demand, v, vii, 1
energy efficiency, 2, 5, 6, 8, 9, 11, 12, 16, 17, 18, 46, 66, 110, 111, 115
energy expenditure, 5
Energy Independence and Security Act, 2
Energy Policy Act of 2005, 2, 10, 45, 51, 52, 70
energy prices, 5, 23
energy supply, 76
England, 12, 72
enrollment, 51, 54, 55, 56, 66, 67, 73, 116

environment(s), 16, 28, 39, 62, 77, 79, 81, 83, 84, 85, 86, 88, 89, 92, 98, 102, 106, 109, 114, 118, 119, 120, 121, 138
environmental conditions, 41
environmental sustainability, 111
equipment, 6, 8, 17, 22, 25, 30, 31, 32, 35, 40, 41, 45, 64, 83, 93, 99, 100, 101, 119, 123, 131
equity, 110
erosion, 68
evidence, 49, 51, 69, 105
evolution, 51, 69
examinations, 79, 92
expenditures, 19, 28, 35
expertise, 29, 44
exposure, 25, 44, 62, 63, 70
externalities, 128, 135

F

field tests, 78, 87
field trials, 37
financial, vii, 3, 6, 12, 20, 28, 30, 41, 43, 55, 80, 84, 93, 95, 101, 108, 109, 112, 116, 117, 121, 128, 134
financial incentives, vii, 3, 84, 121
financial institutions, 41
financial resources, 80, 95, 128, 134
flexibility, 62, 87
food, 41
force, 46, 72
forecasting, 60, 125, 132
formation, 62, 114
forward market, 60
funding, 21, 84, 139
funds, 16, 21

G

global warming, 17
government policy, 10, 16
greenhouse, 17
greenhouse gas, 17
greenhouse gas emissions, 17
grid operators, vii, 3, 32
growth, 10, 12, 16, 23, 43, 53, 57, 66

H

Hawaii, 136
history, 110
host, 93
human, 30, 38, 80, 95
human capital, 80, 95
humidity, 39
hybrid, 22

I

identification, 34, 78, 87
imbalances, 61
IMO, 14, 23
improvements, 6, 8
incompatibility, 35
industrial sectors, 39
industry, 10, 16, 37, 43, 50, 51, 52, 65, 69, 70, 71, 76, 79, 88, 92
information exchange, 32
information technology, 2
infrastructure, 1, 9, 16, 18, 19, 20, 27, 28, 29, 30, 31, 32, 36, 37, 65, 68, 81, 83, 88, 93, 99, 100, 101, 103, 108, 113, 115, 116, 117, 118, 119
institutions, 75, 76
integration, 11, 30, 51, 77, 136, 138
integrators, 35
interface, 33, 36, 109
interoperability, 45
interrelatedness, 80, 93
intervention, 30, 32, 38
investment(s), 5, 9, 16, 17, 18, 19, 20, 27, 28, 29, 49, 81, 83, 84, 93, 99, 100, 103, 108, 110, 111, 113, 115, 116, 118, 119, 121, 127, 134
Iowa, 53
IRC, 76, 97, 137
iron, 57
irrigation, 24

isolation, 104
issues, 65, 78, 80, 83, 85, 88, 91, 93, 95, 98, 99, 112, 117, 119, 121, 136

J

jurisdiction, 81, 95, 96, 114, 118

L

labeling, 12
lack of control, 40
landscape, 29
laws, 11
lead, 4, 8, 9, 27, 29, 107
learning, 49
legislation, 10, 13, 18
lifetime, 123, 131
logistics, 41

M

magnitude, 22, 51, 73, 100, 122, 130
majority, 12, 53, 62, 65, 69, 83, 119
management, vii, 2, 3, 6, 7, 8, 9, 12, 23, 27, 31, 36, 39, 44, 45, 49, 50, 51, 53, 68, 70, 111, 136
manufacturing, 41
market access, 13
market share, 66
market structure, 12, 16, 28, 42
marketing, 8, 19, 57, 65, 68, 74, 115
marketplace, 35, 81, 118
Maryland, 11, 57
mass, 39, 40, 136
materials, 4
MCP, 76, 102, 103, 105
measurement(s), 2, 30, 31, 42, 59, 100, 107, 108
meter, 29, 30, 31, 36, 42, 43, 44, 54, 65, 100, 108, 139
methodology, 28, 43, 73
Mexico, 20
migration, 41, 55
mission, 41
models, 96, 97
modernization, 75
modifications, 21, 65
monopoly, 68, 91, 110
moratorium, 85, 121

N

natural gas, 17
negative consequences, 43
negotiation, 68
network operators, 16, 19, 20
neutral, 24
New England, 4, 10, 11, 12, 27, 45, 46, 54, 57, 68, 72, 76, 96
New South Wales, 21
North America, 16, 20, 70, 72, 76, 89, 95, 137

O

oil, 4
OPA, 2, 15
operations, 29, 30, 41, 44, 68, 75, 79, 86, 96, 99, 100, 109, 112, 113
opportunities, 8, 10, 13, 15, 18, 25, 28, 29, 40, 41, 45, 55, 64, 65, 78, 81, 83, 84, 87, 88, 89, 93, 111, 113, 114, 119, 120, 139
opportunity costs, 16
optimization, 102, 105
organize, 88
outsourcing, 68, 111
overlap, 73
oversight, 114

P

Pacific, 2, 19, 40, 46, 47, 68, 71
pairing, 116
parallel, 18
parity, 19, 28

participants, 6, 12, 13, 14, 29, 32, 33, 42, 54, 56, 57, 58, 59, 65, 69, 73, 93, 103, 104, 114, 124, 131, 132
payback period, 81, 93, 101, 115, 116
peak electrical, vii, 3
penalties, 19, 20, 23, 25, 28, 44, 54, 59, 69, 73
plants, 4, 15, 53, 103
playing, 6, 60, 87
policy, 10, 16, 17, 44, 49, 51, 52, 70, 74, 76, 79, 88, 91, 92
policy choice, 74
policy makers, 76
policymakers, 9, 49, 50, 51, 81, 118, 139
pollution, 4
ponds, 40
portfolio, 1, 16, 18, 25, 30, 44, 55, 58, 69, 77, 84, 85, 91, 98, 103, 104, 139
power generation, 17
power plants, 4, 13, 15
preparation, 40
prevention, 12, 29
price signals, 11, 24, 90
pricing policies, 35
private firms, 8
private sector, 114
procurement, 17, 27, 77, 81, 85, 87, 90, 98, 104, 105, 114, 116, 118
producers, 61
profit, 16, 81, 83, 84, 92, 103, 110, 111, 112, 113, 118, 119, 121, 127, 134
profitability, 19
program administration, 66
project, 61, 73, 76
proposition, 25, 83, 119
public policy, 65, 111
pumps, 39, 40

R

ramp, 13
rate of return, 17, 18, 28, 110, 113
ratepayers, 84, 91, 109, 112, 121, 126, 133
reading, 31
real time, 2, 31, 71
reality, 9
real-time basis, 26
recommendations, 42, 70
recovery, 16, 19, 110, 111, 113, 126, 133
recruiting, 115
regulations, 16, 27
regulatory agencies, 35
regulatory framework, 18, 19
relative size, 104
relief, 19
renewable energy, 17, 77, 85
researchers, 76
reserves, 3, 4, 13, 15, 23, 55, 60, 62, 83, 85, 86, 90, 96, 98, 99, 100, 101, 102, 103, 104, 106, 107, 108, 112, 118, 119, 120, 124, 132, 139
Residential, 24
resource availability, 61
response time, 9, 31, 122, 123, 130, 131
restrictions, 107, 122, 123, 124, 125, 130, 131, 132, 133
restructuring, 50, 111
retirement, 15
revenue, 16, 24, 28, 36, 42, 55, 65, 83, 90, 101, 102, 104, 106, 107, 108, 109, 110, 111, 117, 119, 127, 134
risk(s), 25, 26, 27, 29, 35, 40, 44, 66, 68, 83, 103, 108, 110, 115, 119, 126, 133
rules, 53, 60, 65, 66, 69, 76, 78, 80, 87, 89, 90, 93, 96, 97, 98, 100, 101, 102, 108, 109, 110, 114, 117, 124, 128, 129, 132, 135, 138

S

savings, 5, 7, 30, 63
scarcity, 81, 118
scope, 33, 51, 52, 79, 99, 106, 139
scripts, 32
security, 77, 85, 87, 123, 131
sellers, 90
service provider, 1, 44, 50, 51, 57, 65, 66, 70, 78, 79, 81, 84, 87, 92, 93, 96, 100, 106, 108, 109, 115, 118, 120, 127, 133, 134, 138

shape, 63, 94
shareholders, 91, 110
shelf life, 41
shortfall, 22, 106
signals, vii, 3, 32, 33, 34, 44, 99, 108, 113, 123, 130, 131, 138
society, 85, 121
software, 29, 33, 41, 65, 98, 102
solution, 7
specialists, 43
speed of response, 100
spending, 139
spot market, 23, 90, 92
stability, 60
stakeholders, 34, 35, 36, 43, 49, 51, 79, 92
standardization, 37
state(s), 10, 14, 15, 17, 18, 37, 50, 51, 52, 53, 54, 60, 61, 62, 63, 65, 68, 70, 73, 74, 77, 78, 81, 84, 85, 87, 89, 91, 92, 101, 103, 113, 114, 118, 120, 129, 135, 139
state regulators, 60, 68, 74, 78, 81, 84, 87, 113, 114, 118, 121, 139
statistics, 10
statutes, 81, 87, 118
steel, 4, 53
storage, 39, 40, 75, 76, 96, 97, 122, 123, 124, 125, 126, 127, 128, 129, 130, 131, 132, 133, 134, 135, 136
structure, 10, 15, 16, 21, 22, 33, 50, 80, 90, 91, 95, 109, 128, 129, 135
supplier(s), 12, 83, 84, 91, 92, 102, 107, 108, 119, 129, 139
supply chain, 41
supply curve, 70
sweat, 31, 39
synchronize, 4

T

target, 17, 24, 112, 113
tariff, 19, 24, 25, 65, 74, 77, 80, 84, 86, 89, 95, 96, 106, 109, 113, 121
tariff rates, 89
Task Force, 70
technical assistance, 59, 115
technological advances, 75
technology(s), 6, 8, 21, 29, 30, 31, 32, 33, 35, 37, 41, 42, 45, 65, 66, 68, 74, 80, 81, 83, 99, 100, 111, 115, 116, 118, 119, 128, 134, 139
temperature, 39
Tennessee Valley Authority, 3, 20
territory, 20, 92, 115
testing, 13, 35, 41, 96
time periods, 21, 22
training, 45, 59, 72
transaction costs, 80, 95
transmission, 2, 4, 5, 10, 11, 18, 32, 40, 50, 61, 63, 65, 75, 78, 86, 89, 91, 96, 100, 106, 114, 122, 130, 136
transport, 41
treatment, 10, 11, 19, 28, 40
triggers, 38, 61
TVA, 3, 20, 29

U

U.S. Electricity Markets, v, vii, 46, 49
UK, 13, 16, 19, 20, 21, 23, 28
uniform, 138
United, 10, 13, 19, 45, 49, 51, 52, 68, 69, 77, 91, 92, 138
United Kingdom, 13, 19
United States, 10, 45, 49, 51, 52, 68, 69, 77, 91, 92, 138
universe, 50, 62
up-front costs, 65

V

valuation, 102
variations, 50
ventilation, 2, 39
virtualization, 41
volatility, 51, 107, 114, 124, 131

W

Washington, 46, 70, 71, 72, 136

wastewater, 37
water, 22, 24, 37, 39, 40, 103
water heater, 22, 24
wear, 99
Western Australia, 14, 27
wind power, 60
Wisconsin, 80, 84, 88, 92, 94, 112, 114, 115, 117, 120, 138

Y

yield, 7, 103